Congenital Defects

Holt, Rinehart and Winston
Developmental Biology Series

James D. Ebert
General Editor

LAURI SAXÉN
JUHANI RAPOLA
University of Helsinki, Finland

Congenital Defects

HOLT, RINEHART AND WINSTON, INC.
New York Chicago San Francisco
Atlanta Dallas Montreal
Toronto London Sydney

Library of Congress Catalog Card Number: 76-75926
SBN: 03-079255-X Paper
Printed in the United States of America
1 2 3 4 5 6 7 8 9

Cover: "Nucleolar genes in action." The presence of extrachromosomal nucleoli in amphibian oocytes permits isolation and electron microscopic observation of the genes coding for ribosomal RNA precursor molecules. Visualization of these genes is possible because many precursor molecules are simultaneously synthesized on each gene. Individual active genes are separated by "spacers," stretches of DNA that are apparently inactive at the time of ribosomal RNA synthesis. Electron micrograph by O. L. Miller, Jr. and Barbara R. Beatty, Biology Division, Oak Ridge National Laboratory.

Preface

The science of congenital defects can be considered a typical interdisciplinary field. Based on normal genetics and embryology, it represents a branch of pure biology; in exploring the etiology and pathogenesis of abnormal conditions it is closely related to general pathology; and, finally, insofar as it is concerned with serious structural and functional disorders, it might be regarded as a field of clinical medicine. A book on congenital defects cannot hope to cover all these different aspects of the subject, therefore, a suitable compromise has to be made in presenting the information gained from experimental studies and clinical investigations. The authors of this book have felt that the field could be treated as a branch of developmental biology and at the same time be extended to include the valuable data obtained by people working in the medical profession. In many contexts, the best examples to illustrate different defects and their pathogenesis were afforded by clinical medicine, since, after all, man is in several respects the best studied of all species. In their presentation of these, the authors have intentionally omitted the use of medical terminology in the hope that the reader who is not familiar with this language will profit from the data obtained outside the strict scope of developmental biology. Readers belonging to the medical profession, on the other hand, must excuse us for not having differentiated between observations made on *E. coli,* the frog, the mouse, or man, provided the information related to our problem.

Due to the interdisciplinary nature of the field to be covered in this book, the authors have greatly profited from consultations with colleagues in different fields of biology and medicine. We would, therefore, like to express our deep gratitude to all of them and particularly to our friends Kari Cantell, H. R. Nevanlinna, Eero Saksela, Erkki Saxén, and Sulo Toivonen. Moreover, we would like to extend thanks to our consulting editor, Dr. James Ebert, for pleasant collaboration and many valuable suggestions. Those who provided pictorial material for this book have greatly contributed to its completion and we only hope that they will feel that their material has been adequately used.

All drawings have been prepared by Mrs. Lorna Lucy Tallberg, who has also assisted us with the typescript throughout its preparation. We express our deep gratitude to her as well as to Mrs. Jean Margaret Perttunen for her expert correction of our English language. The references have been checked by Mrs. Marita Antila, to whom we are greatly indebted.

Finally, we express the hope that this book will stimulate some of our younger colleagues to focus their interest on the field of teratology, which certainly offers exciting theoretical problems to be explored by the employment of all the means afforded by the rapid advances in basic biology. An applied science like teratology must always lag behind the achievements of the pure sciences on which it is built, but it is our feeling that modern genetics and developmental biology have now reached a stage at which their information can be profitably applied to the study of abnormal embryogenesis.

Urajärvi, Finland L.S.

J.R.

Contents

Preface v

1 *THE PROBLEM AND THE TASK* 1
Definition 2
Incidence of Congenital Defects 3

2 *METHODS IN TERATOLOGY* 7
Nonexperimental Methods 7
Experimental Teratology 19

3 *GENETIC ASPECTS OF CONGENITAL DEFECTS* 35
Simple Inheritance of Congenital Defects 35
Multifactorial Etiology of Hereditary Defects 46
Chromosomal Abnormalities 57
Maternal-Fetal Incompatibility 71

4 *GENESIS OF CONGENITAL DEFECTS* 78
Morphogenetic Movements 78
Tissue Interactions 85
Growth 97
Degeneration 104

5 *SENSITIVE PERIODS IN DEVELOPMENT* 112
Gametopathies 114
Blastopathies 116
Embryopathies 118
Factors Determining Sensitive Periods 126
Fetopathies 134

6 *ENDOCRINES AND MALDEVELOPMENT* 140
Ontogeny of the Endocrine Organs and Hormonal Function 141
Placenta and Feto-Placental Unit 141
The Effect of Hormones on Differentiation and Development 144
Birth Defects Caused by Endocrinologic Factors 148
Diabetes Mellitus 155
Hormones as Teratogens 161

7 *RADIATION HAZARDS* 165
Radiation Injury on the Cellular Level 166
The Effect of Irradiation on the Embryo 171
Effect of Radiation on Human Development 177

8 *CHEMICAL TERATOGENESIS* 182
General Principles 183
Thalidomide Embryopathy 199
History of Meclizine 205

9 *VIRUS AND EMBRYO* 210
Virus-Cell Interaction 210
Viral Susceptibility of Embryonic Cells 211
Malformations Experimentally Induced by Viruses 215
Rubella 217
Virus Infection as a Cause of Human Maldevelopment 222
Factors Affecting Viral Susceptibility 227
Viral Lesions 235

Index 239

1

The Problem and the Task

Gross structural defects, monstrosities, detectable at birth have been known and described as curiosities from the most remote times and we find them reproduced in ancient Egyptian paintings and prehistoric Peruvian pottery. Until the beginning of this century they were considered merely as rarities without real scientific or practical significance and discussions on their etiology were mainly concerned with divine will, heavenly bodies, demons, and the evil eye. No doubt, our ideas of these disorders are still confused by similar or related beliefs, but at least we are beginning to realize the magnitude of the problem both in clinical practice and in basic biological research. The study of congenital defects, *teratology,* has grown into a fascinating field of biology and offers an approach to the understanding not only of abnormal development, but of normal developmental events as well. In this introductory chapter, we shall outline the field and its problems and evaluate its "practical" and "theoretical" significance in modern biology and medicine.

Whereas very many severe disorders of early childhood can nowadays be either prevented or effectively treated, very little progress has been made in the prevention of congenital defects. As an example of the development during this century, the relative importance of two major causes of neonatal death are compared in Fig. 1.1. Consequently, these defects have become one of the leading causes of death during late intrauterine life and account for one of the largest groups of patients in the pediatric wards—calculations made in the United States and in the Scandinavian countries have indicated that no less than one-third of the capacity of these wards is devoted to the treat-

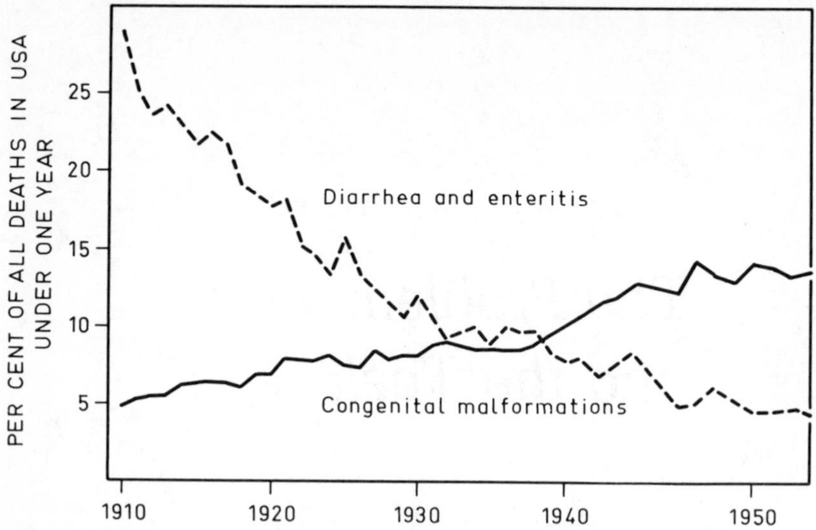

Fig. 1.1 Deaths from diarrhea and enteritis compared to those from congenital malformations in children under one year of age. (From J. Warkany. Pediatrics, **19**, Suppl. 725 [1957].)

ment of congenital defects and their sequelae. Therefore, teratologic research has at least two highly important practical objectives: to provide information on the etiology of the defects in order to contribute to their prevention and to bring to light facts on which adequate therapy can be based.

DEFINITION

Instead of using the classic terms congenital "malformations" or "anomalies," we prefer to speak of congenital "defects." This naturally expands the topic, since "malformations" would refer merely to structural aberrations, whereas "defects" includes all abnormalities at the morphological and biochemical levels *before or at birth.* For operational convenience, however, the definition, although theoretically acceptable, should be limited to defects or disorders detectable at birth or, if detected later, definitely attributable to antenatal development uninfluenced by subsequent factors. For there is a wide variety of abnormal conditions and diseases detectable during late postnatal life for which a congenital background, genetic or nongenetic, may be suggested, but where the final manifestation might be the consequence of a series of subsequent events and exposures to exogenous factors. Certain neoplastic diseases appearing decades later may be mentioned as typical examples. Here scientists often suggest congenital factors (genetic predisposition, intrauterine "initiation" of the disease, or true embryonal "rests" later giving rise to the neoplasm). Until the etiology of these and related diseases is fully clarified, the "embryonic" background should not be taken for granted and, hence, we shall exclude them from our definition of congenital "defects."

What then is meant by the "abnormalities" included in our definition? We should be fully aware of how greatly meanings may differ in this respect and how much such differences of interpretation can confuse the statistics of congenital defects to be presented below. A pigmented nevus of the skin may not be a normal structural constituent and yet its high incidence seems to exclude it from being regarded as "abnormal" or "defective." Furthermore, another benign neoplasm (or defect), the capillary hemangioma, is a relatively common, harmless, and often regressive lesion but may in a few cases constitute a true and serious congenital defect. Should it therefore always be considered a defect or should its size and location decide whether we include it in our statistics? We could list a great variety of similar intermediate conditions, lesions, and abnormalities for which no definition can provide an answer, so we will stress only the existence of the problem and keep it in mind when analyzing and comparing different statistics of the incidence of "congenital defects."

INCIDENCE OF CONGENITAL DEFECTS

Incidence figures for congenital defects (often only malformations) are available for the common laboratory rodents and especially for human populations. The overall incidence of malformations in the statistics for the latter is of the order of 1 to 3 percent (see Kennedy, 1967), but the figures are subject to many sources of errors which render them relatively unreliable if not useless. We may briefly mention some of these.

Detection Rate

Most of the extensive statistics are based on observations made soon after birth, usually in the maternity ward. Consequently, such surveys overlook a great variety of congenital defects and disorders not detectable during those early days of life, among which we may mention mental retardation, defects of the sense organs, abnormal sexual development and certain enzyme defects known as "inborn errors of metabolism." The low detection rate at birth can be clearly seen in Table 1.1, where the results of reexaminations of series of children studied soon after birth are presented. We see that not more than 50 percent of the defects were detected during the first few days of life. We may note, too, that the "final" incidence figures vary greatly, which is mainly due to the difficulty of defining "congenital defects" and to individual differences in the recording of these.

TABLE 1.1
Follow-up Studies on the Incidence of Congenital Defects[a]

Number of Children	Frequency at Birth	Observation Time	Follow-up (percent)	Frequency ("Final")
1. 672	1.5	4 years	75	10.1
2. 3179	2.4	3–6 years	91	16
3. 56760	1.73	5 years	95	2.31
4. 5531	3.5	6 months	89	6.0
		12 months	85	6.9
5. 63796	1.02	9 months	25	3.12

[a] Compiled from the literature as follows:
1. V. P. Coffey and W. J. E. Jessop: Lancet **i**:748, 1963.
2. A. D. McDonald: Brit. J. Prev. Soc. Med. **15**:154, 1961.
3. T. McKeown and R. G. Record: Ciba Found. Symp. on Congenital Malformations, p. 2. Churchill, London, 1960.
4. G. W. Mellin: Birth Defects, p. 1. Lippincott, Philadelphia, 1963.
5. J. W. Neel: Am. J. Hum. Genet. **10**:398, 1958.

Lost Cases

It has been calculated that at least 20 percent of fertilized human zygotes are lost during early pregnancy (see Carr, 1967 and Inhorn, 1967) and these would consequently escape all statistics on congenital defects based on children born during the third trimester. It is extremely difficult to estimate the role of developmental defects in these early fetal deaths. In studies on spontaneous abortions, analyzable embryonic material is obtained only in relatively few cases and abnormalities have been detected in 20–40 percent of these cases (see Javert, 1957, Sentrakul, and Potter, 1966). The causative role of developmental defects in the remaining 75 percent of the series cannot be judged. (The very same difficulty is encountered by scientists in experimental teratology, as will be discussed in Chapter 2.) The above estimations may add altogether some 5 percent to the overall figures for the incidence of defects in fertilized eggs.

Stillbirths

Fetal deaths occurring during late pregnancy constitute another factor limiting evaluation of the incidence of congenital defects. Only an autopsy performed by a skilled pathologist can give us reliable information on this point, thus permitting an estimate of the role of developmental defects in late fetal deaths. But even here, not all the stillborn children can be analyzed, as a great many of them are severely macerated and cannot be examined in detail. Autopsy series of stillbirths have produced fairly uniform results and in eight independent series compiled from the literature the incidences of defects were as follows: 16.0, 20.4, 25.0, 22.5, 23.6, 15.9, 13.3, and 19.5 (see Klemetti, 1966). In field studies based on nonautopsied material most of these 20 percent would have been lost and such statistics are therefore of little or no value.

Conclusions

On the basis of the studies briefly mentioned above and our personal experience, we may summarize our knowledge of the incidence of congenital defects in human embryos and children in Table 1.2.

TABLE 1.2
Estimates of the Incidence of Congenital Defects in Human Embryos, Stillborn Children, and Liveborn Individuals.

Early Fetal Deaths			
Total incidence of abortions	20%		
embryonic defects	25%	total incidence	5%
Detected at Birth			
minor defects	1%		
major defects	0.5%		
lethal defects	0.5%	total incidence	2%
Detected in Follow-up Studies			3%
		total incidence	10%

SUGGESTED READINGS

Carr, D. H. Cytogenetics of abortions. *In* Comparative Aspects of Reproductive Failure, (K. Benirschke, ed.), pp. 96–117. Springer-Verlag, New York (1967).

Inhorn, S. L. Chromosomal studies of spontaneous human abortions. *In* Advances in Teratology, (D. H. M. Woollam, ed.), vol. II, pp. 37–99. Logos Press, Acad. Press, London (1967).

Javert, C. T. Spontaneous and Habitual Abortion. McGraw-Hill, New York (1957).

Kennedy, W. P. Epidemiologic aspects of the problem of congenital malformations. Birth Defects Original Article Series, (D. Bergsma, ed.), vol. III, no. **2**: 1–18 (1967).

Klemetti, A. Relationship of selected environmental factors to pregnancy outcome and congenital malformations. Ann. Paediat. Fenn. 12, Suppl. **26**: 1–71 (1966).

Sentrakul, P. and E. L. Potter. Pathologic diagnosis on 2,681 abortions at the Chicago Lying-in Hospital, 1957–1965. Am. J. Pbl. Hlth. **56**: 2083–2092 (1966).

2

Methods in Teratology

The achievements in a certain field of biology or medicine are often a reflection of the development of the methods available to the scientist. Thus it is always important to be aware of the methodological history underlying our knowledge. Moreover, when evaluating the present state of knowledge in a certain field, the student should know how this information has been acquired and what sources of error, pitfalls and biases may have affected the original observations that form the basis of the generalizations appearing in textbooks. As regards teratology, there are very few methods originally applied to this field. Use has been made of methods developed by epidemiologists, geneticists and embryologists, and the suitability of these methods for studying the etiology and mechanism of maldevelopment may often be questionable.

For the above reasons, it is necessary to devote a whole chapter to acquainting the reader with the possibilities, fallacies, and limitations of the methods available to the teratologist today. Some methods and their limitations have already been presented in Chapter 1 and further examples will be dealt with throughout this book. Here, only the general principles underlying the different approaches will be discussed, together with some illustrative examples.

NONEXPERIMENTAL METHODS

The incidence of malformations in children being on the order of 3 to 5 percent (Chapter 1), an immense body of data has accumulated and is continuing to accumulate which is at the disposal of scientists searching for the causative factors underlying maldevelopment. Hence, these data should be effectively and critically analyzed, although it should be emphasized that the limitations and fallacies of all clinico-epidemiologic studies are marked and that comparatively little information has so far been derived from such investigations concerning the etiology of maldevelopment.

Registers and Community Surveillance

Since the main task of epidemiologic studies is the detection of causal associations between etiologic factors and their consequences as well as the statistical verification of these correlations, ample data are required. Registers compiled from different sources and planned community surveillances may be useful in offering such ample material for statistics. At the same time, however, they are subject to a variety of sources of error, some of which will be discussed below.

Detection rate As stated in Chapter 1, not more thant 50 percent of congenital defects are detectable at birth and yet most of the statistics available are based on information collected from lying-in hospitals or during early postnatal life. It is, therefore, hardly legitimate to speak of the "incidence" of defects in a certain population and the term "detection rate" appears more adequate. For the same reason, valid comparisons of statistics from different sources and different countries are very difficult to make and in addition to the "detection rate," the definition of malformations and congenital defects varies greatly. However, this does not necessarily imply that such methods and selected materials should not be used in the search for etiologic factors in maldevelopment.

Lost cases With a few exceptions all statistics on congenital defects are based on information collected from children born during the last trimester of pregnancy. Consequently, no data are available on defects in embryos lost during early pregnancy and yet there is good reason to believe that the damage caused by teratogens is frequently severe enough to bring about the death of the embryo. (Experimental embryology can provide numerous examples of this.) In fact, it is not inconceivable that this loss of study material might not only obscure an effect, but might lead to an actually erroneous result, where a teratogen has eliminated the weak embryos appearing in the control group as malformed children. Such a situation would be extremely difficult to demonstrate, but unfortunately equally difficult to exclude in clinico-epidemiologic studies in which reliable information on early abortions very seldom is available.

Sampling and randomization All epidemiologic studies are based on analyses of samples and their ultimate task is to arrive at generalizations applicable to the population represented by the sample. A sample must fulfill two conditions:

1. It should be representative of the population about which inferences are to be drawn.
2. The data collected must be adequate to yield an answer to the questions posed.

Regarding epidemiologic studies in teratology, there is every reason to accept these two basic conditions and to stress their importance, which seems all too often to be forgotten. Sampling and its randomization is a question comprehensively discussed in textbooks of statistics, and here we need only quote the statement made by Yule and Kendall (1940), that "the human being is an extremely poor instrument for the conduct of random selection." The other condition mentioned above raises the question of the magnitude of the material required for teratologic studies. Owing to the limiting factors discussed in this and other chapters most epidemiologic studies have yielded negative results as far as the demonstration of a causal relationship between an exogenous teratogen and abnormal development is concerned. The only conclusion to be drawn from such investigations is that "no causal relationship could be detected," which, however, is no proof of lack of causal association. Only very seldom have the investigators bothered about the significance of a negative finding, that is, the acceptance of the null hypothesis (no correlation exists), and tested the power of their method. The few cases where this has been done have yielded rather depressing results.

Taylor (1960) refers to a study in which the incidence of cerebral palsy was studied in children of mothers with nonpuerperal complications of pregnancy as compared with those of mothers without such complications. Approximately 1 percent of the mothers had suffered from such complications and the incidence of cerebral palsy among their children was 18.2 per 1000, as compared with 5 per 1000 in the controls. Taylor could show that this trebling of the rate had an almost 50 percent chance to be missed in his population of 40,000! If this probability of missing a true effect is to be reduced to the usually accepted 5 percent level, an incidence rate of 8 times that in the control population is required. In another study, use was made of a mass vaccination which could be timed exactly and during which approximately 20 percent of women in child-bearing age were vaccinated during a period of one month (Saxén *et al.*, 1968). Subsequent analysis of its effect on the birth rate gave no evidence of a decrease in the number of livebirths 8 to 9 months later, which might be taken to indicate that no increased rate of abortions had resulted from the vaccination. However, analysis of the power of the method showed that in this case only an abortion rate of at least 50 percent would have been detected at an acceptable level of significance in the population under study, with its 25,000 annual births.

Controls One of the main difficulties in all epidemiologic studies is the choice of an adequate control series. If the starting point is children with congenital defects, an acceptable control can be collected from the children born immediately after or prior to the defective ones in the same hospital or maternal welfare district, or delivered by the same physician. Selective factors, such as geographic, racial and economic differences, are in this way eliminated and there is at least good reason to believe that no parameters other than the outcome of pregnancy have been employed in selecting the study and the control group. (Another problem is the analysis of the possible differences in the maternal history, as will be shown below.) If, on the other hand, the material has been selected according to exposure to a certain teratogen, the choice of controls is more difficult. If selection has been based on the consumption of certain drugs, would it be adequate to study a control where this drug has not been used? Or would it not be more appropriate to search for mothers with the same basic disorder but not treated with the drug under investigation? And because it is often difficult to find untreated cases, could a series of patients treated with another drug be used as a control? If so, what factors have determined the choice between these different drugs and could these selective factors (social class, allergy, and so on) have had a bearing on the development of the embryo? Most of these and similar questions remain unanswered in teratologic studies, but the investigator should always be aware of the extreme complexity of the situation when data on human beings is submitted to clinico-epidemiologic analysis and when a control series has to be chosen.

Detector defects In view of the multiple etiology of malformations, the overall incidence (or detection rate) of congenital defects may not always be useful in detecting specific causative factors. In cases where a teratogen results in a variety of malformations, the overall incidence is naturally the only point of reference, but it may be expected that in most cases a slight increase in some malformations will be masked by the permanent "background" caused by very many other teratogens. Hence our hope has to be based on a more specific effect, detectable as an increase or difference in a certain type of malformation (the effect of thalidomide would never have been recog-

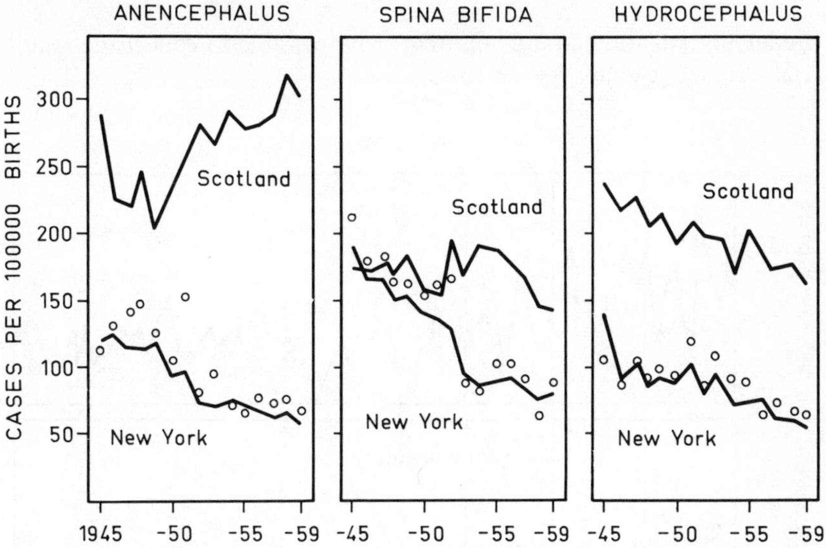

Fig. 2.1 Incidence of certain defects of the central nervous system in Scotland and New York (o: New York first births). (After T. H. Ingalls and M. A. Klingberg. Am. J. Med. Sci. **249:** 316 [1965].)

nized as an increase in the overall incidence of congenital defects). On the other hand, the low incidence of each specific defect makes it very difficult to obtain series large enough for statistical treatment and instead there is a tendency to analyze groups of malformations seemingly representing a uniform entity. Even this may be unjustified, as illustrated in Figs. 2.1 and 2.2. Malformations of the central nervous system are often treated as one group, but even the examples in the figures are sufficient to indicate that this is unjustified in a search for specific causative agents. The differences, both in the seasonal variation and in the general trend, between the two different series strongly suggest that different factors must be responsible for the various types of defects in the central nervous system and that combining these might, in fact, obscure the true differences observable in epidemiologic studies.

The above considerations have led teratologists to use certain "markers" in their research on the factors responsible for maldevelopment. From the point of view of the scientist searching for causative agents in teratology, the significance of such detectors bears no relation to their clinical or social importance. The important points are that the marker defect be a definite anatomic or metabolic entity, that the diagnosis be reliable, that the diagnostic possibilities determining the detection rate be equally distributed among the population under investigation, and finally that it be possible to collect the material without selective loss of cases.

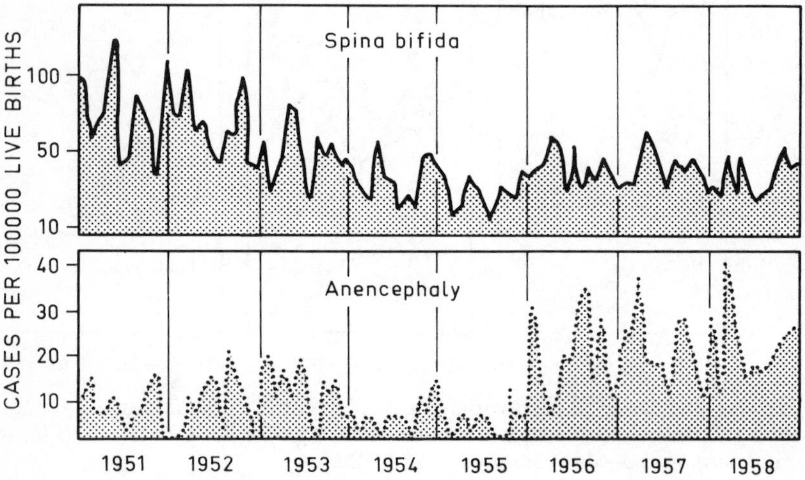

Fig. 2.2 Incidence of spina bifida and anencephaly by month in Pennsylvania. (After J. G. Babbot and T. H. Ingalls. Am. J. Public Health **52:** 2009 [1962].)

What to look for Epidemiologic studies on human populations have revealed a great number of marked differences and changes in the incidence of various malformations in different populations and at different periods. As an example, some studies on the incidence of anencephaly will be briefly discussed, as this malformation may be considered to fulfill the requirements of a good "marker" listed above. This anomaly is easily detectable, the detection rate hardly being affected by improvement of or differences in diagnostic possibilities and, hence, even statistics from different countries are likely to be comparable. Changes in the incidence of anencephaly in a defined population have already been indicated in Figs. 2.1 and 2.2 but the most dramatic change was reported in postwar Berlin in 1945–1948 (Fig. 2.3). This rapid change can hardly be explained by changes in the genetic material of the populations and strongly suggests that environmental factors (nutrition, diseases, and so forth) play a role. The high incidence of this defect in Scotland has led to several detailed studies demonstrating a definite seasonal variation in the monthly incidence of anencephaly. It is always tempting to correlate such variations with possible teratogenic factors acting at

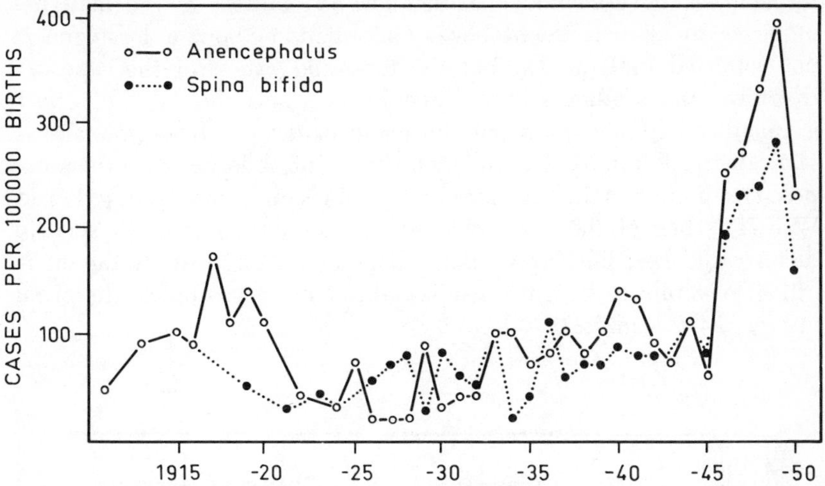

Fig. 2.3 Incidence of spina bifida and anencephaly in Berlin in 1915–1950. (After H. Gesenius. Intern. J. Sexol. **6:** 24 [1962].)

the time of conception and during early pregnancy. It is easy to suggest such factors. In the present example close correlations to sunshine and temperature were noted and, naturally, a great many other seasonal factors would equally well fit the case. However, it is questionable if such apparent coincidences are conclusive at all. On the other hand, such data might be valuable for excluding environmental factors suggested to be teratogenic.

All the above data might lead the reader to adopt a rather pessimistic attitude toward all clinico-epidemiologic investigations, and it seems, in fact, that not too much should be expected from them until we know what kind of factors to look for. In the future, however, in combination with experience gained from experimental teratology and hints gleaned in these studies, the information obtained from epidemiologic surveys may prove extremely useful.

Specific Opportunities for Clinico-Epidemiologic Studies

What has been said above on the difficulties of extracting specific causative factors from the general milieu does not apply in certain exceptional situations. Epidemics and pandemics, sudden changes in nutritional factors, exposure of large populations to radiation, and so on, offer some kind of natural experimental conditions suitable for analysis by epidemiologic methods. In contrast to the maternal inquiries to be considered below, information about a teratogen is not collected individually, but the time and extent of the exposure is known from other sources (see below) and the harmful effect is calculated from register data on malformations or from preplanned clinical investigations. The follow-up study of children of mothers exposed to radiation in the nuclear bombings in Japan during World War II is one of the first epidemiologic studies in which a specific factor could be analyzed in a large population. To illustrate the interesting possibilities of such studies, three more examples are given in Figs. 2.4, 2.5, and 2.6.

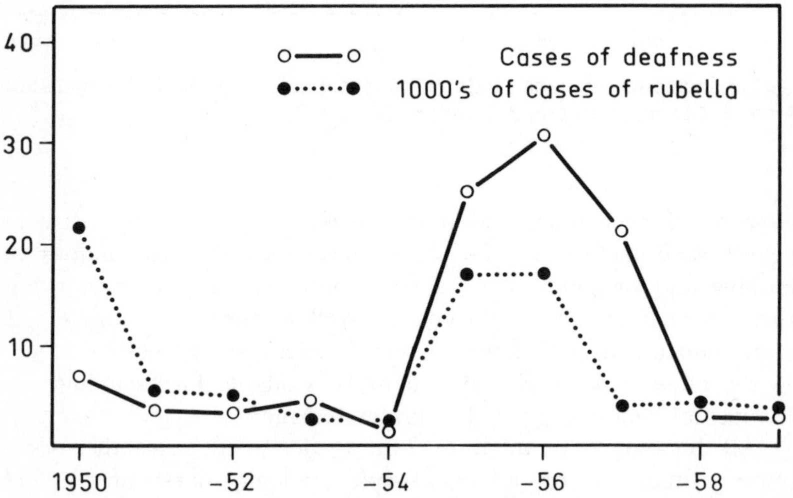

Fig. 2.4 Correlations between the annual incidence of rubella and congenital deafness (Canada). (After P. Statten and M. Wood. Proc. Cand. Otol. Soc. [1961].)

Rubella embryopathy was detected as long ago as 1941 (see Chapter 9) and the epidemiologic study illustrated in Fig. 2.4 does not add any essentially new information to the problem. Knowing that an infection during early fetal life may cause, among other disturbances, lesions of the inner ear leading to congenital deafness, the authors investigated the annual incidences of reported cases of rubella in a certain district in relation to the annual rates of deafness and showed a definite correlation. A similar study performed by Lentz (1965) indicated a close correlation between the wholesale figures for thalidomide and severe limb malformations of the phocomelia-

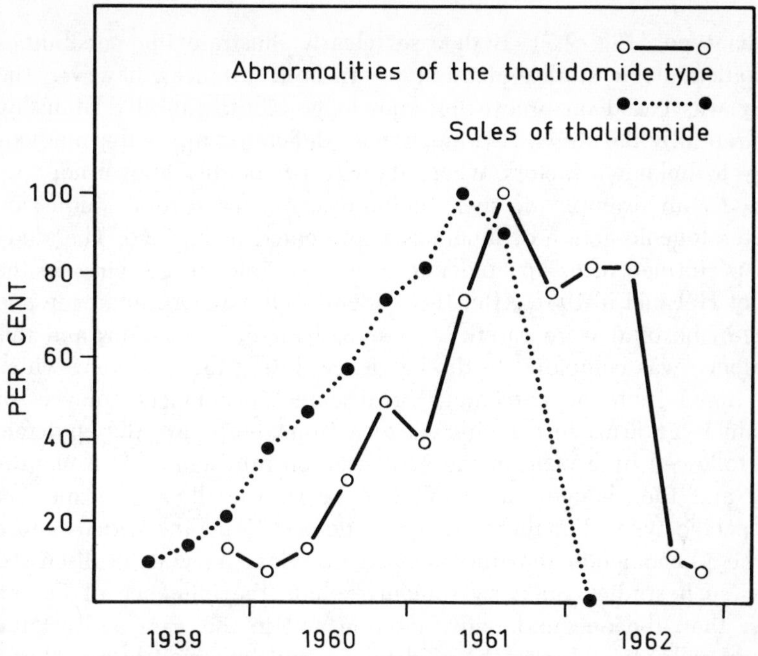

Fig. 2.5 Correlations of the wholesome figures of the drug thalidomide to the incidence of severe limb malformations (Germany). (After W. Lenz. *In* Embryopathic activity of drugs, J. M. Robson, F. M. Sullivan and R. L. Smith [eds], p. 182, Churchill, London [1965].)

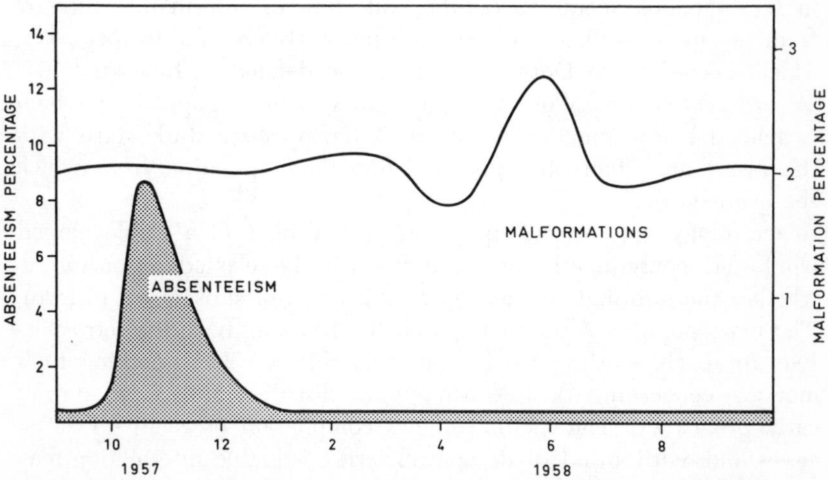

Fig. 2.6 Rise in the incidence of malformations subsequent to an epidemic caused by influenza-A virus (Helsinki). (After L. Saxen *et al.* Acta Pathol. Microbiol. Scand. **49:** 114 [1960].)

amelia type (Fig. 2.5). Both cases clearly illustrate the possibilities of epidemiologic studies of this type. In both instances, however, the study was based on preexisting knowledge of the causal relationship illustrated by the study. It is much more difficult to apply this methodology to unknown factors, where the effect is neither known nor specific. As an example of such limitations, one of several studies on the teratogenic action of influenza is presented in Fig. 2.6. The study was performed after an epidemic caused by influenza-A virus in the city of Helsinki in 1957. After the epidemic, all newborn infants in one lying-in hospital were carefully examined for malformations and this incidence was compared to the incidence during the epidemic, which was timed from the percentage of absenteeism among employees of certain large firms in the city. As seen from the figure, the epidemic was followed by a peak in the malformation rate some 7 to 8 months later and the change was verified to be statistically significant. Yet no specific type of malformation was detected and the actual nature of the phenomenon remains an open question. As will be discussed later, such studies can very seldom exclude the influence of factors other than the one under investigation and in this case at least the heavy medication during the epidemic should be seriously considered before the observed effect is attributed to the virus.

Prospective Versus Retrospective Inquiries

A seemingly easy approach to the problem of causative agents in teratogenesis seems to be the collection of information directly from the mothers. This can be done prospectively or retrospectively, which, according to Dorn (1955), can be defined as follows:
"A *prospective* or follow-up study starts with a population which is defined before some event occurs. A *retrospective* study starts with the event and then attempts to define the population from which the event arose."
In teratology, this "event" apparently is the birth of a maldeveloped child and consequently these studies can be classed according to whether the information was collected prior to or subsequent to birth. The more popular of the two approaches has long been the retrospective approach, starting from anomalous children and querying their mothers concerning drug consumption, diseases, and such, during early pregnancy. The method allows compilation of relatively large series and, with an adequate control series, valuable information may be obtained. The main source of error here is the maternal memory bias. The inquiry deals with complications and events which took place some eight to ten months earlier, and this interval is long enough

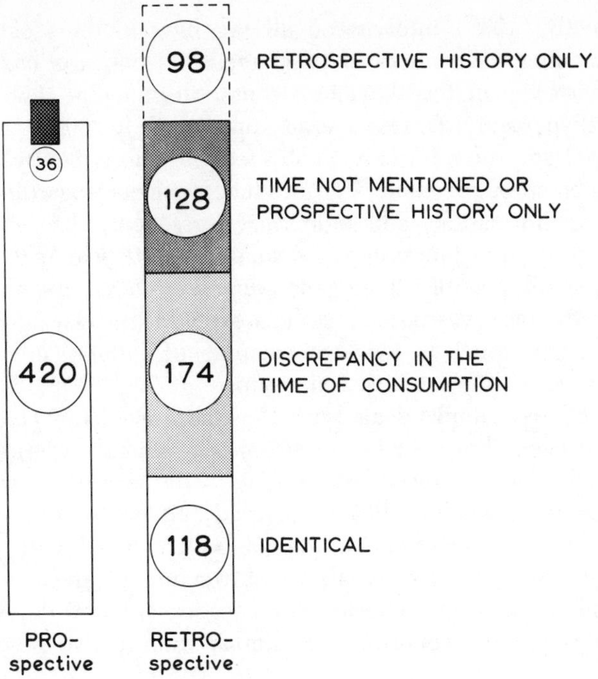

Fig. 2.7 Comparison of prospective and retrospective inquiries on consumption of drugs in early pregnancy. (From A. Klemetti and L. Saxén. Am. J. Public Health **57:** 2071 [1967].)

to produce rather unreliable replies, as seen in Fig. 2.7. The results shown in the figure were obtained in a prospective study in which approximately 4000 mothers were carefully questioned during early pregnancy (fifth month). After delivery, the mothers of malformed or stillborn children were requestioned, together with a control group of the same size, and the replies were individually compared. The results show that only one-fifth of the replies retrospectively collected tallied fully with the prospective data. Moreover, the memory bias may be selective, as mothers of malformed children are inclined to find a reason for the maldevelopment, whereas mothers of healthy babies are no longer interested in complications of early pregnancy.

A prospective study seems to exclude this memory bias and at least its selective effect can be avoided. The great disadvantage of such studies is the enormous series that has to be collected before the number of malformations is sufficient for statistical treatment: for every detectable malformation, approximately 30 mothers subsequently giving birth to healthy children must be interviewed. Two such studies might be briefly described.

Klemetti (1966) interviewed all pregnant mothers of a certain district in their fifth month of pregnancy and only the mothers who did not visit the maternal welfare center during that time were excluded (25 percent). During 1 year, approximately 4000 mothers were interviewed and the information collected was subsequently compared to the outcome of pregnancy. To produce evidence regarding specific types of malformations and individual teratogens, the series would have had to be split into groups too small for statistical analysis. However, when all possible teratogens were considered together, a significant difference was noted: mothers exposed to diseases, drugs and irradiation gave birth to twice the number of stillborn or malformed children as those of the unexposed group.

Our other example deals with the drug meclizine (meclozine), which has been shown to be teratogenic in animal experiments and suggested in some retrospective studies to be harmful to the human embryo as well. Mellin (1963) recorded the use of this drug prospectively in 3200 pregnancies, 266 of which ended with the birth of a malformed child. Comparison of the use of the drug in this group and in the control series did not disclose any differences—a result subsequently confirmed in similar prospective studies (see Chapter 8).

Controlled Retrospective Surveys

Both prospective and retrospective inquiries are based on information given by the mother and are consequently affected not only by the memory bias, but also by limitations due to her insufficient knowledge. She may not be able to tell the exact names, formulas or dosages of drugs used, nor be aware of the true nature of her illness or of the diagnostic procedure employed. In addition, there are the subclinical inapparent diseases and infections known in some cases to be teratogenic (rubella)—all of which will be definitely lost from an inquiry. As an example, we may mention a study performed in Baltimore after the Asian influenza epidemic. When maternal information regarding the influenza was compared to serologic evidence of recent influenza infection, the overlap was only 25 percent. Such disadvantages can be excluded only by combining different methods; a prospective inquiry may be combined with the compilation of all relevant information obtainable from physicians' prescriptions, hospital records, and so on, and simultaneous serologic studies, like those of Brown, referred to on page 223. In special cases, limited approaches may be useful, as in the recent study of Bunde and Leyland (1965), where a method was developed to collect retrospective information on the consumption of a certain drug. In cooperation with physicians using these drugs in their practices, a study group was collected and compared with an adequate control group of women who had not taken the compound. Here, however, we come up against the most confusing disadvantage in all epidemiologic studies, the complexity of the experimental situation.

An experimental teratologist working in strict laboratory conditions is still worried about the unknown and confusing factors possibly influencing his preplanned experiments, where usually only one variable is unknown and under investigation. An epidemiologist, analyzing causal relationships in an experiment performed by *nature* or by the *populace,* seldom meets conditions where only one factor distinguishes his study group from the controls. A careful maternal history, large series and repeated studies may slowly indicate one of the intermingled factors to be a teratogen, and subsequent animal experiments may confirm this suspicion.

EXPERIMENTAL TERATOLOGY

Animal Experiments

Much of our present knowledge of the mechanism of maldevelopment and the mode of action of genetic and nongenetic teratogenic factors is based on studies with common laboratory animals. In addition to such basic studies, animal experiments have become a most important screening method in the pharmaceutic industry, and as far as new drugs are concerned, they offer the only practical possibility of testing the teratologic side effects. In the succeeding pages, some general methods used in present-day experimental teratology will be presented and stress will be laid on a variety of factors to be considered in planning and performing such investigations. Most of these factors can be crystallized in what is known as the "teratologist's creed": In order to produce congenital defects, one must have the TTSD of teratology, that is: appropriate *time,* appropriate *treatment,* appropriate *susceptibility* and appropriate *dose* (see Fraser, 1964). In addition to the presentation that follows, the reader is referred to Chapter 5, where the time factor is discussed extensively. In this presentation of the different methods of experimental teratology, only very few examples of the results obtained will be illustrated here, but complete results may be found in other chapters of this volume.

Test animals Experimental embryology, the basis of today's teratology, was largely founded on studies of lower vertebrates, especially amphibian and avian embryos. The profound knowledge of their normal development, the relative ease of operational manipulations of these embryos, and the possibility of direct application of the teratogen (the role of the maternal organism being excluded during treatment), made these embryos very useful in experimental work. Perhaps the most popular among these subjects has been the chick embryo, and the papers by Klein (1965) and by Karnofsky (1965) give useful information regarding performance of such experiments. No results of such experimental investigations on lower vertebrates will be presented here, but the reader is again referred to other chapters, especially Chapter 4, where the mechanism of abnormal development is discussed.

In mammalian teratology, all of the common laboratory animals have been employed, although most of our present information comes from experiments on mice and rats. Recently, several investigators have used monkeys with promising results. Yet it is impossible to outline any general rules to follow in making the choice of the most suitable experimental animal and it should be born in mind that different species seem to vary greatly in their susceptibility to teratogens: cortisone induces cleft palate in mice and rabbits, but not in rats, and experiments with thalidomide have also yielded variable results. Not only does the susceptibility vary greatly among different species, but also between different strains of the same species, as has been found by many investigators. This difference in susceptibility between different mouse strains is illustrated in Table 2.1, where the effect of cortisone on the incidence of cleft palate has been analyzed (see also Chapters 3 and 4 and the recent reviews by Dagg, 1966, and Smithberg, 1967).

TABLE **2.1**
Incidence of Cortisone-induced Cleft Palate in Different Strains of Inbred Mice.[a]

Strain	Percent
CBA	12
C57BL	19
C3H	68
DBA	92
A	100

[a] From H. Kalter, in Teratology. Principles and Techniques, (J. G. Wilson and J. Warkany, eds.), University of Chicago Press, Chicago 1965.

These and many similar strain-specific differences in chemical (as well as viral and radiational) susceptibility often make it very difficult to compare results obtained in different laboratories and the use of random-bred strains immediately invalidates many comparisons. On the other hand, it cannot be denied that such a random strain might, in fact, be of some advantage in teratologic studies. Finally, it should be stressed that the use of tissue fragments or cells of different strains does not necessarily exclude misinterpretations due to strain-specific differences in susceptibility. Both viral resistance and differences in drug sensitivity have often been shown to be maintained in tissue culture conditions.

Experimental procedure Figure 2.8 gives the results of an animal experiment which can be considered one of the classic approaches in basic teratology performed in 1953 by Wilson, Roth, and Warkany.

Let us now briefly review the different steps of such experiments and illustrate some of the points to be taken into consideration. Depending on the type of teratogen or teratogenic conditions, the treat-

Fig. 2.8 The effect of vitamin A treatment on different days of pregnancy in mothers kept on a vitamin-deficient diet. The results show the "sensitive" period for vitamin A deficiency and for the different organs affected. (Based on data from J. G. Wilson *et al.* Am. J. Anat. **92:** 189 [1953].)

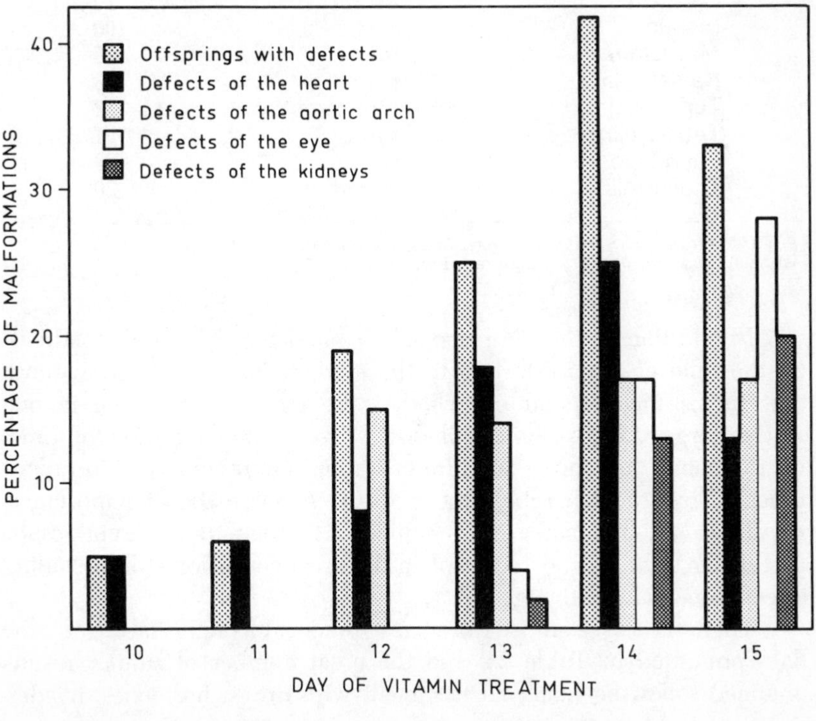

ment can be made prior to pregnancy or at different stages of development of the embryo. Certain "chronic" conditions are produced prior to mating (hypovitaminosis, fasting, thyroidectomy, and so forth) but usually the teratogen to be investigated must be given during the proper sensitive period (see Chapter 5). Neither the sensitive period nor the dosage for a particular teratogen can usually be predicted and, hence, both should be varied in order to gain a true idea of its effect.

Some idea of the great variability of the teratogenic dose of drugs can be obtained from Table 2.2 where the effective teratogenic dose (on a kilogram ratio basis) is compared to the therapeutic dose in man.

TABLE **2.2**
The Ratio Between the Teratogenic Dose of Certain Drugs in Animal Experiments and the Therapeutic Dose in Man.[a]

Drug	Experimental Animal	Ratio
Caffeine	mouse	35
Cortisone	rat	—
Cortisone	mouse	400
Cortisone	rabbit	20
Insulin	rat	100
Meclizine	rat	35
Na salicylate	rat	3
Terramycin	mouse	12
Tetracycline	rat	1
Thyroxine	rat	2
Thalidomide	rabbit	20

[a] From C. Fraser, in Second Internat. Confer. Cong. Malformat., (M. Fishbein, ed.,) Internat. Med. Congress Ltd., New York, 1964.

In addition to the dose dependence and the definite dependence on the time of administration to the mother, there is much evidence that the results depend on whether the teratogen is given in one or two large doses or in small doses over a longer period of time. Furthermore, the mode of administration (intravenously, intraperitoneally, by stomach tube, and so forth) seems to be of importance. All these variables make it very difficult to compare different results and this may to some extent explain the great confusion still prevailing in experimental teratology.

There is no need to draw any practical conclusions from the data presented in Table 2.2 and the great number of similar results obtained following maternal treatment with drugs, hormones, irradiation or viruses. One point, however, should be stressed. In many instances the effect of the teratogen on the mother herself has been forgotten and the factor has been considered to be directly teratogenic athough it is evident that the embryos suffer in the first place from the severe disturbances of the maternal organism. As an example, tetracycline in high doses can be mentioned. This drug leads to fetal death and resorption, but at the same time severe lesions with liver necrosis can be observed in the mother and it is not possible to decide which was the primary effect. The same reasoning applies in cases of maternal hypoxia, irradiation, severe stress, virus infections, and so on. It is generally believed that the fetal organism is always more sensitive to toxic factors, but the opposite may be true in some cases and this should be taken into consideration.

Finally, a great many environmental factors require consideration and control during an experiment. The season is known to affect the results, the age and parity of the mother; her nutritional condition can change the susceptibility of the embryos; and latent diseases of the experimental animals may have a potentiating effect—to mention only a few variabilities. Errors arising from all these sources can be excluded only by the use of a proper control group of the same strain, age, and size, kept in exactly similar conditions and treated identically with the experimental group—except in respect to one factor. As mentioned before, the possibility to do this is the great advantage of experimental teratology as compared with the complex situation in epidemiologic studies. The investigator, therefore, should make full use of it.

Examination of the pregnancy outcome The purpose of examining the outcome of pregnancy after a given treatment is quite simply to register all deviations from normal (that is, from the controls). To avoid the cannibalism frequently occurring among laboratory rodents, the fetuses should be collected and analyzed before birth—the nineteenth day of pregnancy is frequently suggested for mice and rats. Different types of protocol sheets have been proposed by different authors for the recording of pregnancy outcome; the reader will find an example in the paper of Wilson (1965).

The first observation to be made is the number and site of fetuses in the uterus. In addition it is important to record the number of resorbed embryos or implanation sites (see Fig. 2.9).

After the size and viability of the fetuses have been recorded, they are exposed to detailed individual analysis. Depending on the problem to be studied, all possible histologic, histochemical, and microchemical analytic methods, as well as immunologic and radiochemical techniques can be used. There is no need to discuss them

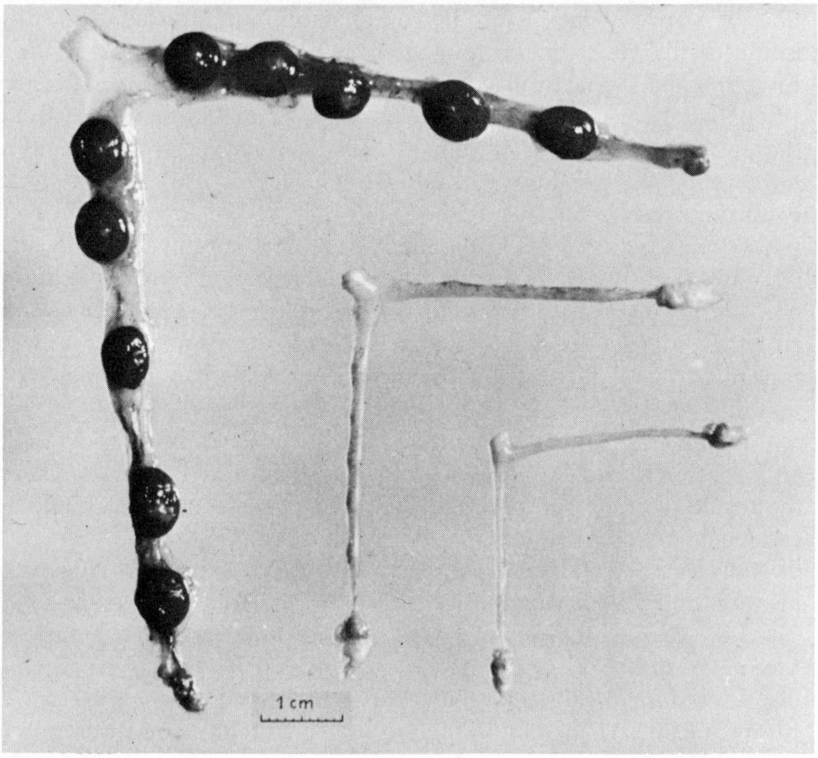

Fig. 2.9 (A) Mouse uterus with normal placentas on day 19 (fetuses removed). (B) Uterus on the same day showing original implantation sites after abortion on day 11. (C) Uterus of a nonpregnant mouse. (From H. Frohberg and H. Oettel. Ind. Med. Surg. **35:** 113 [1966].)

here in detail, but two methods primarily intended for teratologic studies will be described. Both of these have repeatedly been proposed for use in large screening series and in combination they give rather a good general idea of the stage of development and possible malformations at the morphologic level.

The screening method developed by Wilson (see Wilson, 1965) is merely a thorough serial analysis of the embryo. After fixation, the embryos are gently decalcified and cut by hand with a razor blade into sections 1 to 2 mm thick. These can then be analyzed under the dissecting microscope and if deviations from normal are observed, histologic sections can be prepared from the sites of interest. For the analysis of the skeleton, the claring procedure with subsequent alizarin staining of the mineralized parts has proven to be more reliable than x-ray examination (Fig. 2.10).

Increasing evidence is accumulating that certain treatments dur-

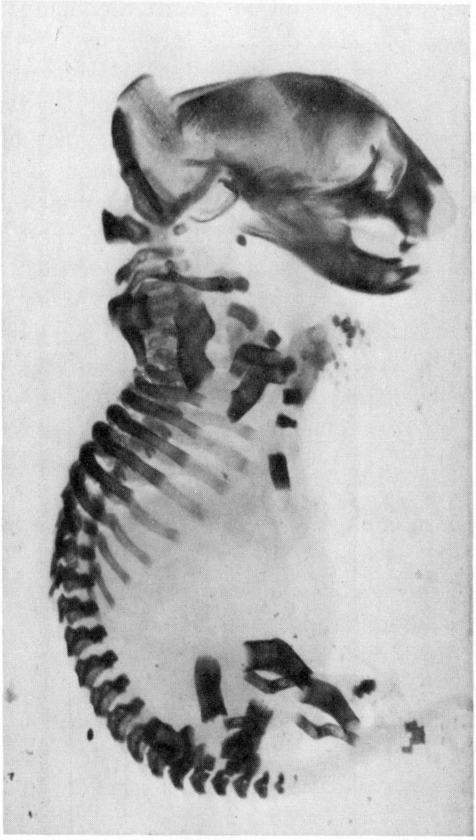

Fig. 2.10 The skeleton of an 18-day-old mouse embryo cleared with KOH and stained with alizarin. (Courtesy of Dr. Juhani Kohonen.)

ing pregnancy may lead—as might be expected—to defects in the central nervous system not detectable as morphologic malformations, but observable as abnormal behavioral development of the experimental animals. These observations demand the extension of teratologic analysis to cover the perinatal period and the use of methods developed by experimental psychology (p. 177). Similar late manifestations to be considered and analyzed during postnatal life of embryos treated *in utero* might be a general growth retardation, late manifestation of virus infection (oncogenic viruses), disturbances in endocrine function, and shortening of the life span. In conclusion, the teratologist should not be satisfied with positive (or negative) findings in fully developed fetuses, but should pay serious attention to teratogenic "markers" detectable only during postnatal life.

Finally, a comment should be made on the lost embryos. As

stressed, it is always important to record the number and sites of resorbed embryos, but this does not give any information on the nature of the lesion. For this purpose both blastocysts and early embryos have been successfully examined. Lutwak-Mann and Hay (1962) have developed a technique for preparing flat-mounts of rabbit blastocysts and have reported interesting changes due to drug consumption by the mother (changes in size and mitotic index, degenerative changes, and so forth). Ferm (1965), on his part, has developed a rapid method by which pregnant hamsters were treated with various teratogens on the eighth day of gestation and the embryos were analyzed 24 hours later. After application of known teratogens, severe malformations were noted and, as the author stresses, many of these would not have been detected in routine examinations of advanced stages because by that time the embryos would already have been resorbed.

Intrauterine manipulations Some recent experiments in which embryos have been treated locally *in utero* may prove very important. As stated before, the age of the embryos and the exact time of conception are difficult to determine and, hence, the best control might be to use embryos in the same uterus (the other horn) as the controls to treated ones. The uterine vascular clamping method developed in rats by Brent (1965) seems to afford a possibility of such controlled experiments. Short-time clamping of the uterine vessels (in fact of the whole uterine horn) usually does not lead to high mortality and the fetuses will be of the same size as those in the nonclamped control horn. The embryos temporarily cut off from the general circulation in this way provide good material for testing different teratogens to which the controls will not be exposed. The same author has also developed techniques for intrauterine irradiation, where one horn is shielded and serves as a control. As shown in the earlier works by Wolff (1963) on chick embryos, part of a single embryo can here be exposed to radiation and the rest shielded. Modern surgical and anesthesiologic techniques allow a rather complicated manipulation—clamping plus partial radiation—without loss of embryos. Another example of sophisticated intrauterine operations is afforded by the transplantation experiments of Schinckel and Ferguson (1963). Since these skin graftings on fetal lambs, similar experiments have been performed successfully on monkeys and offer a straightforward method for studying the development of the immune system.

Finally, we may summarize some of the most obvious disadvantages of animal experiments, which make it impossible to extrapolate results from species to species or from animal to man and which make it very difficult to compare results obtained in different laboratories.

1. Susceptibility of different species and strains varies greatly.
2. Condition of the mother and a variety of environmental factors influence the results (nutrition, season, age, and so forth).
3. The period of susceptibility is often short and variable and the exact stage of development at treatment is difficult to determine.
4. Development of the embryos cannot be followed continuously, and minor effects (growth retardation and so on) are difficult to detect. Reversible effects cannot be studied.

Many of these limitations and sources of errors can be excluded by careful planning of the experiments, followed by detailed analysis of the results and by choice of an adequate control series. Nevertheless, new methods are still needed and, in particular, the possibility of performing teratologic experiments in tissue culture conditions should be mentioned.

In Vitro Experiments

There are basically two types of culture in which animal cells can be kept alive outside the organism. They may be grown either as single cells in monolayers, invading a spongeous material or suspended free in a liquid medium, or as organized tissues in organ cultures. The former, here referred to as the "cell culture," usually leads to progressive loss of phenotypic characteristics (dedifferentiation) and consequently takes us further away from the original *in vivo* situation. These cells may be regarded as prototype cells and they have been used extensively in biologic studies, where cell responses have been analyzed on the cellular and subcellular levels. These cells may in the future be most useful for analyzing the effect of teratogenic agents, the mechanism of their action, and the processes leading to the expression of an abnormal or defective genome. Many of the fundamentals of virology and radiobiology to be referred to in this book are based on studies of cell cultures; but teratology in the narrow sense has so far been mainly interested in targets where the original differentiative stage and tissue pattern have been preserved. These conditions are almost met in "organ cultures" where whole embryonic organs or fragments of developing tissue are maintained *in vitro*. Some of the possibilities and advantages, as well as limitations of this method, will be discussed below.

Figure 2.11 gives an idea of the method and the two most fre-

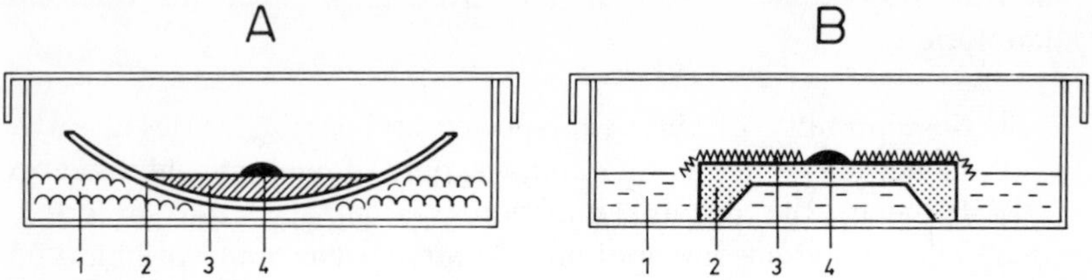

Fig. 2.11 Two types of set-ups for organ culture. (A) Method developed by Fell where the tissue (4) lies on a clotted medium (3) on a watch-glass (2). The bottom of the dish is covered with cotton-wool (1). (B) A Trowell-type culture technique where the tissue is placed on a strip of filter paper (3) on a metal screen (2) and these are immersed in liquid culture medium (1).

quently used modifications of it. The piece of tissue, preferably not more than 2 × 2 × 2 mm in size, is placed at the medium gas interface and the dish is usually kept in a 5 percent CO_2 atmosphere at 37°C. Numerous culture media have been described for different purposes. Most of them contain a certain amount of serum and embryo extract, which inevitably brings a number of unknown variables into the picture. It is, therefore, of great interest that several tissue rudiments have been cultivated for prolonged periods in "chemically defined" media, the constituents of which are known precisely (see Biggers *et al.*, 1961, Wolff *et al.*, 1963).

Some of the advantages of such methods are illustrated by the experiment schematically demonstrated in Fig. 2.12. The possible effect of tetracycline on the development and calcification of embryonic bones was studied by cultivating the rudiments in a chemically defined medium. The teratogen was added to the medium in various concentrations and its effect was followed by different parameters (length of the calcified zone, uptake of radiocalcium and increase of the total amount of calcium). In such experiments, the course of development and any deviations that occur can be followed continuously (see below). Further, the experiment can be stopped at any time and the medium changed to one containing no teratogen. In addition, the stage of development of the organ at the start of the experiment can be determined with great accuracy and an adequate control can be obtained. (In the case of paired organs the most appropriate control is naturally the untreated contralateral one.) Some of the further advantages of such experiments will be discussed below.

As we have already seen in this chapter, and as will be noted repeatedly in the course of this book, cleft palate has become one

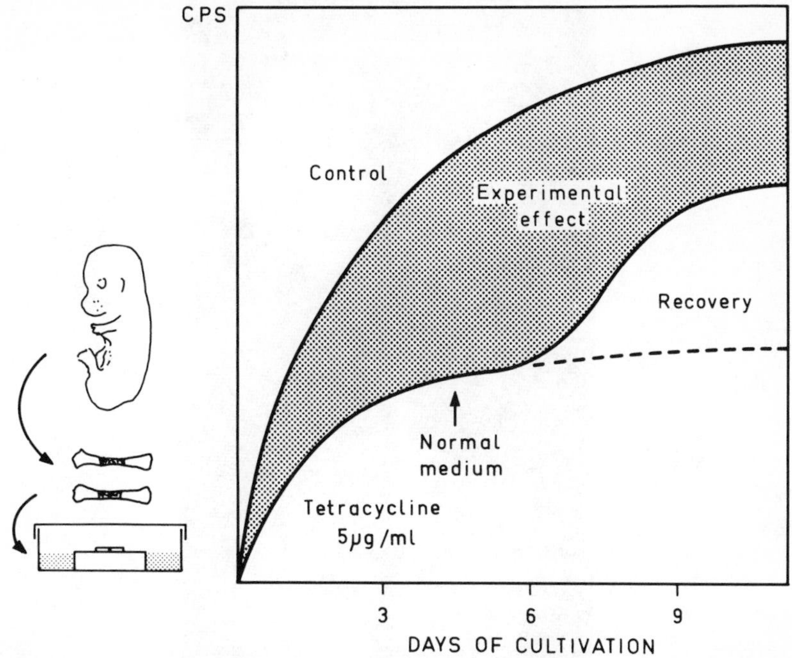

Fig. 2.12 A general scheme and some results of experiments in which the effect of tetracycline on osteogenesis has been tested in organ cultures. The bone rudiments removed from 16-day-old embryos were first cultivated in the presence of tetracycline, and a definite decrease in the uptake of radiocalcium was noted as compared to control cultures. On day 4 *in vitro* the bones were transferred to normal culture medium, and after a short lag period an increased uptake (partial recovery) was observed. (After L. Saxén, J. Exp. Zool. **162:** 269 [1966].)

of the most valuable teratogenic "markers" in both experimental and epidemiologic studies. It is, therefore, understandable that the development of the palate has also been studied in controllable *in vitro* conditions—and with considerable success. Palatal processes of both mouse and rat embryos regularly fuse *in vitro,* giving rise to a morphologically normal secondary palate (Fig. 2.13) although development takes place at a slower rate than *in vivo* (see Konegni *et al.,* 1965). This model system appears to offer excellent opportunities for studying the effect of different teratogenic compounds and especially the strain-specific differences in susceptibility known from *in vivo* experiments.

At the beginning of the paragraph on experimental teratology, we stressed the possibilities offered by lower vertebrates as targets for operative manipulations under controlled conditions. The develop-

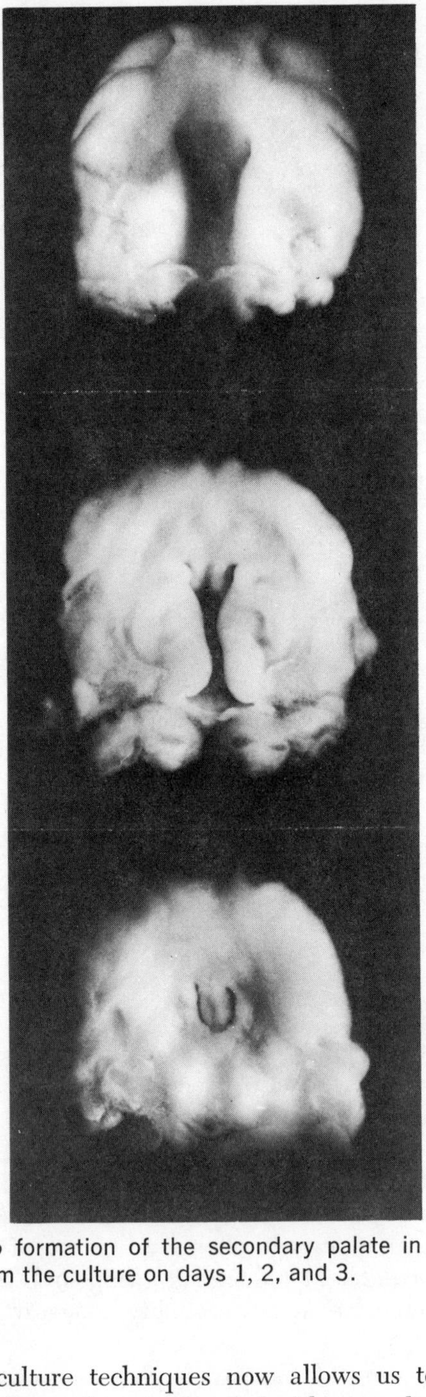

Fig. 2.13 *In vitro* formation of the secondary palate in mouse embryos. Samples taken from the culture on days 1, 2, and 3.

ment of organ culture techniques now allows us to perform such experiments on mammalian tissues and either to alter the course of development or to construct model systems where the development

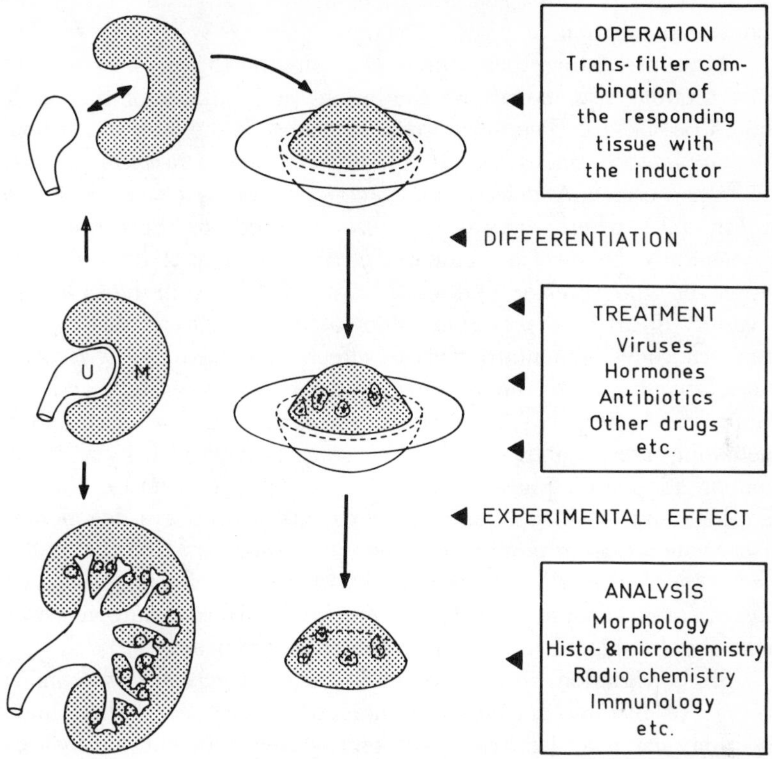

Fig. 2.14 Model system to study the induction and development of kidney secretory tubules developed by Grobstein (1956). After removal of the natural inducer of tubules (the ureteric bud), the metanephrogenic mesenchyme is placed on a Millipore filter and an inductor tissue (spinal cord) is cemented beneath. During subsequent development, the tissue can be treated with different teratogens at various times and the effect can be read from the mesenchyme by different analytical methods. The time required for an inductive stimulus to exert its effects is of the order of 24 hours, after which the mesenchyme will differentiate autonomously and be treated and analyzed separately.

of different tissue components and their interactive processes can be followed. From the point of view of experimental teratology, one additional advantage is the possibility of using known triggers of differentiation (inductors), and consequently being able to start a certain developmental process at any given time with subsequent analysis of the sequential differentiation and its divergences. One such model is illustrated in Fig. 2.14 and the results obtained by the employment of this system will be referred to in connection with the discussion of the developmental mechanisms underlying maldevelopment (page

39), sensitive periods (page 128), chemical teratogenesis (page 198), and the acquisition of virus resistance (page 213).

After presenting these three examples of the possibilities of *in vitro* teratology, some of the limitations and failings of the method should be stressed. The most obvious of these is the fact that although development continues to rather advanced stages *in vitro*, it is still far from normal. A definite delay can always be observed, as seen in Fig. 2.13, where some 4 days are required for the formation of a secondary palate (as compared with less than 2 days *in vivo*). Moreover, the transfer and cultivation of tissues *in vitro* leads to a variety of artificial processes which should be strictly distinguished from true developmental events or effects produced by the teratogen under investigation: the *in vitro* transfer is always followed by a short "shock period," in which some of the cells are lost and the metabolic rate declines rapidly. This stage is succeeded by a "restoration or adaptation phase," which is very easy to confuse with true developmental events. Finally, the explants will always degenerate—after some weeks or months, depending on the tissue and the methods used—and this must not be confused with specific effects of the chemicals, viruses or other teratogens employed. Here again, a parallel control culture is the only reliable point of reference.

Slight variations in culture conditions (pH, gaseous environment, change of medium, mechanical manipulation of the dishes, and so on) may have a decisive effect on the development and viability of the explants *in vitro* and, hence, before any teratologic studies can be performed, the culture conditions should be carefully standardized and checked. What we have said about environmental factors in animal experiments is even more important here and many observations *in vitro* may, in fact, be true artifacts dating from unknown factors and poorly standardized conditions. The complexity and significance of environmental factors in the experimental situation *in vitro* may finally be exemplified by the analysis of muscle development illustrated in Fig. 2.15.

To study the role of the tension caused by increasing distance between the insertion and origin of a muscle *in vivo*, Nakai (1965) built an experimental model system to imitate this normal tension. Intercostal muscles of chick embryos were fixed on a metal mesh *in vitro* and one end of the explant was connected with silver wires to a rotating motor, causing a permanent pull on the muscle (roughly corresponding to that caused by the increasing intercostal space *in vivo*). A control muscle mounted on the same grid but not pulled served as the control. The results clearly demonstrated the morphogenetic significance of tension: the pulled muscle developed and differentiated well, whereas the control showed poor differentiation and signs of degeneration.

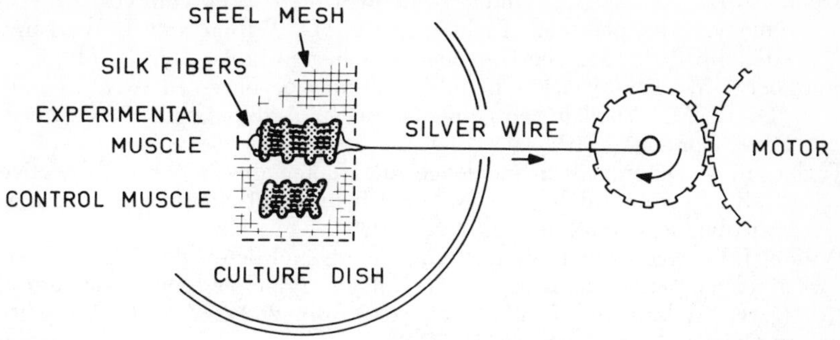

Fig. 2.15 The experimental set-up for the analysis of the role of tension in muscle differentiation *in vitro.* (After J. Nakai. Exp. Cell Res. **40:** 3074 [1965].)

Despite the limitations mentioned above, the sensitive methods of tissue culture may become a valuable addition to our poor repertoire of experimental investigations to elucidate the etiology and mechanism of abnormal development.

SUGGESTED READINGS

Review Papers

Brent, R. L. Uterine vascular clumping and other surgical techniques in experimental mammalian embryology. *In* Teratology, Principles and Techniques, (J. G. Wilson and J. Warkany, eds.), pp. 233–250. The University of Chicago Press, Chicago (1965).

Dagg, C. P. Teratogenesis. *In* Biology of the Laboratory Mouse, (E. L. Green, ed.), pp. 309–328. McGraw-Hill Co., New York (1966).

Doll, R. Interpretations of epidemiologic data. Cancer Res. **23:** 1613–1623 (1963).

Dorn, H. F. Some applications of biometry in the collection and evaluation of medical data. J. Chronic Diseases **1:** 638–664 (1955).

Edwards, J. H. The epidemiology of congenital malformations. *In* Second International Conference of Congenital Malformations, (M. Fischbein, ed.) pp. 297–305. The International Medical Congress Ltd., New York (1964).

Fraser, F. C. Causes of congenital malformations in human beings. J. Chronic Diseases **10:** 97–110 (1959).

———. Experimental teratogenesis in relation to congenital malformations in man. *In* Second International Conference of Congenital Malformations (M. Fischbein, ed.) pp. 277–287. The International Medical Congress Ltd., New York (1964).

Karnofsky, D. A. Mechanism of action of certain growth-inhibiting drugs. *In* Teratology, Principles and Techniques, (J. G. Wilson and J. Warkany, eds.) pp. 185–213. The University of Chicago Press, Chicago (1965).

Klein, N. H. Teratological studies with explanted chick embryos. *In* Teratology, Principles and Techniques. (J. G. Wilson and J. Warkany, eds.) pp. 131–141. The University of Chicago Press, Chicago (1965).

Smithberg, M. Teratogenesis in inbred strains of mice. *In* Adv. Teratol. (D. H. M. Woollam, ed.) vol. II, pp. 257–288. Logos Press, Acad. Press, London (1967).

Taylor, R. J. Differences in incidence rates detectable in the collaborative study of cerebral palsy, etc. National Institute of Neurological Diseases and Blindness, Washington, D.C. (1960).

Wilson, J. G. Methods for administering agents and detecting malformations in experimental animals. *In* Teratology, Principles and Techniques, (J. G. Wilson and J. Warkany, eds.), pp. 262–277. The University of Chicago Press, Chicago (1965).

Yule, G. U. and M. G. Kendall. An Introduction to the Theory of Statistics. Griffin, London (1940).

Special Articles

Biggers, J. D., R. B. L. Gwatkin, and S. Heyner. The growth of avian and mammalian tibiae on a relatively simple chemically defined medium. Exp. Cell Res. **25**: 41–58 (1961).

Bunde, C. A. and H. M. Leyland. A controlled retrospective survey in evaluation of teratogenicity. J. New Drugs **5**: 193–198 (1965).

Ferm, V. H. The rapid detection of teratogenic activity. Lab. Invest. **14**: 1500–1505 (1965).

Grobstein, C. Trans-filter induction of tubules in mouse metanephrogenic mesenchyme. Exp. Cell Res. **10**: 424–440 (1956).

Klemetti, A. Relationship of Selected Environmental Factors to Pregnancy Outcome and Congenital Malformations. Ann. Paediat. Fenniae, **12**: Suppl. No. 26, (1966).

Konegni, J. S., B. C. Chan, T. M. Moriarty, S. Weinstein, and R. D. Gibson. A comparison of standard organ culture and standard transplant techniques in the fusion of the palatal processes of rat embryos. Cleft Palate J. **2**: 219–228 (1965).

Lutwak-Mann, C. and M. F. Hay. Effect on the early embryo of agents administered in the mother. Brit. Med. J. **ii**: 944–946 (1962).

Mellin, G. W. Fetal life study. A prospective epidemiologic study of prenatal influences on fetal development. Meclozine and other drugs. Symposium international de Bruxelles, 1963. Bull. Soc. Roy. Belge Gynécol. Obstet. **33**: 79–86 (1963).

Saxén, L., K. Cantell, and M. Hakama. Relationship between smallpox vaccination and the outcome of pregnancy. Am. J. Public Health **58**: 1910–1921 (1968).

Schinckel, P. G. and K. A. Ferguson. Skin transplantation in the fetal lamb. Australian J. Biol. Sci. **6**: 533–546 (1963).

Wolff, Et., K. Haffen, et Em. Wolff. Les besoins nutritifs des organes sexués embryonnaires en culture *in vitro*. Ann. Nutr. Aliment. **7**: 5–22 (1963).

3

Genetic Aspects of Congenital Defects

Two views of the etiology of congenital defects have struggled for supremacy in the past and it has been vigorously debated whether the majority of the defects were attributable to genetic or extrinsic factors. In many cases there is ample evidence, both experimental and clinical, indicating one or the other of these factors to be the sole cause of a certain malformation. However, it may be roughly estimated that not more than 20 percent of all congenital defects in man could be explained on the basis of simple dominant or recessive inheritance with full penetration, while some 10 percent may be due to the known viral infections or other extrinsic factors and another 10 percent may be counted on major chromosomal abnormalities (Neel, 1961). Yet in the great majority of the defects no single cause is demonstrable. It may be supposed that in many of the remaining cases the etiology is multifactorial, the defect being the outcome of the interaction of several genetic or genetic and epigenetic factors.

The genetic factors producing congenital defects can be divided into different categories, as in Table 3.1. It is not our purpose to review details of general or medical genetics, but to concentrate on a few topics which we feel are pertinent to an understanding of the etiology and pathogenesis of congenital defects.

SIMPLE INHERITANCE OF CONGENITAL DEFECTS

Several malformations inherited as autosomal dominants have been described in the medical literature. Brachydactyly (short

TABLE 3.1 Categories of Genetic Factors.

Single Mutant Gene	
	Autosomal dominant inheritance
	X-linked dominant inheritance
	X-linked recessive inheritance in XY individuals
Pair of Mutant Alleles	
	Autosomal recessive inheritance
	X-linked recessive inheritance in XX individuals
Polygenic Inheritance:	
	Several genes in combination produce the defect or modify the expression of the mutant gene.
	Inheritance of differential susceptibility to exogenous teratogens.
	Gross chromosomal abnormality causes defects as a result of imbalance of the genetic material in the individual.
	Inherited characteristics cause incompatibility between mother and embryo, although both are genetically healthy in the general sense.

fingers), first described by Farabee in the beginning of this century and since then followed through seven generations, is one of the best known examples (Fig. 3.1).

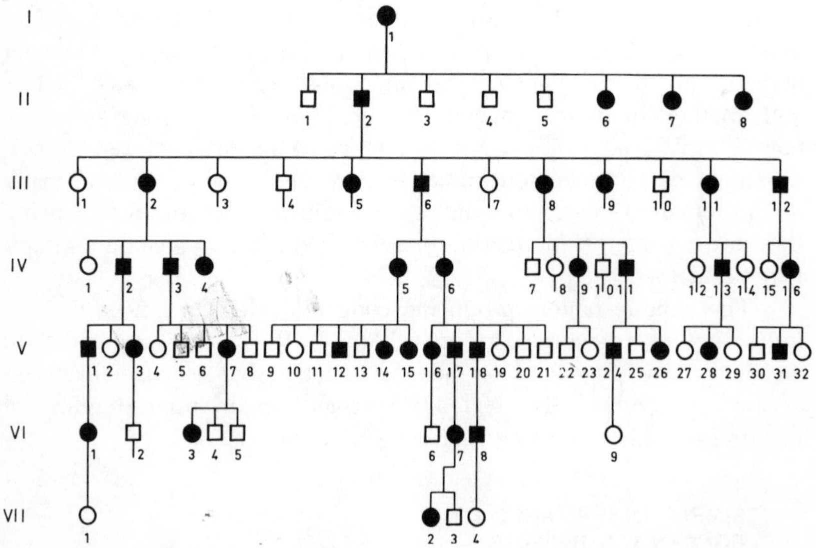

Fig. 3.1 Pedigree of a brachydactylous family through seven generations. (From V. A. McKusick. Human Genetics, Prentice-Hall, Englewood Cliffs, New Jersey [1965].)

Congenital defects caused by dominant genes are usually mild if they survive in the general population. This is due mainly to natural selection. Any defect which prevents or lessens the reproductive capacity of the affected individual is likely to be selected out of the population. The same does not hold true of the rare defects caused by recessive genes because the heterozygotes are of normal phenotype and the mutant gene can be spread by them in the population. In simple autosomal recessive inheritance, the gene, to be expressed in the offspring, must come from both parents. In the mating of two heterozygotes only one-fourth of the offspring will be affected, two-fourths will be carriers of the gene and one-fourth will be both phenotypically and genotypically normal. Related individuals are more likely to be carriers of the same mutant gene and hence carriers of rare recessive genes are often found among the offspring of related parents. The rarer the mutant, the lower is the proportion of homozygotes whose parents are unrelated. Marriages between relatives are frequent in certain social and ethnic groups and in geographical isolates; these marriages have provided good opportunities for genetic studies of recessive inheritance in man. Congenital nephrosis is an example of a rare disease recessively inherited and usually found in geographically isolated communities. This disorder, due to pathologic changes in nephrons, is characterized by generalized oedema as a result of hypoproteinemia and proteinuria. It terminates fatally during the first few years of life (Fig. 3.2).

The genes of the X-chromosome, which may be dominant or recessive, behave in the female like autosomal genes, with certain exceptions due to the inactivation or "Lyonization" of one of the X-chromosomes in the cells of the female (see page 70). The male, with a single X-chromosome, is always hemizygous. As a consequence, the dominance versus recessivity of his genes can be judged only on the basis of studies on females. A male gene carrier cannot transmit the gene to any of his sons, because none of them receives the X-chromosome from him, but he transmits the gene to all of his daughters. A classic example of recessive X-linked inheritance is the hemophilia in the European Royal Families, which is traceable to Queen Victoria of England. A typical pedigree of hemophilia A is shown in Fig. 3.3.

Expressivity and Penetrance

The manifestation of a gene is not always an all-or-none phenomenon; the affected individuals exhibit the particular defect in different degrees of severity, often being intermediately affected. This is particularly true of the dominantly inherited genetic defects. Sometimes the expressivity is much reduced, the overlap with the general

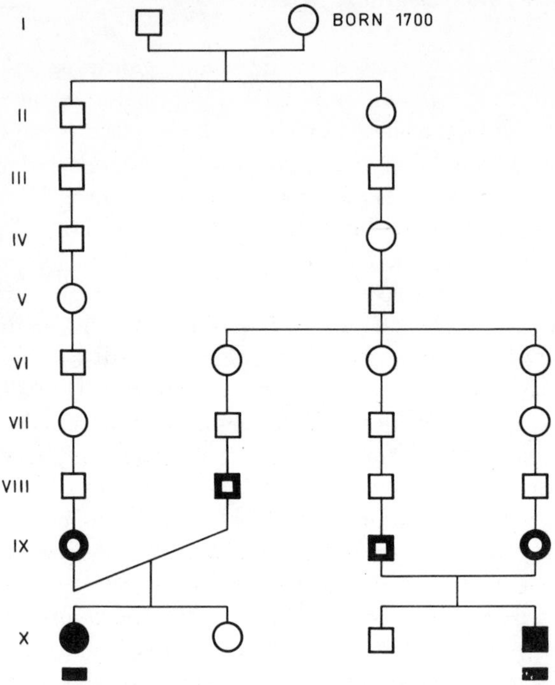

Fig. 3.2 Simplified pedigree of congenital nephrosis in two families. Probands can be traced back to common ancestors, all inhabitants of a relatively small, sparsely populated and isolated area in Finland. The photograph presents the typical appearance of a 6-month-old patient suffering from the disease. (After R. Norio. Ann. Paediat. Fenn. **12,** Suppl. 27, [1966] and from N. Hallman *et al.* Acta Paediat. Scand. Suppl. 172, [1967].)

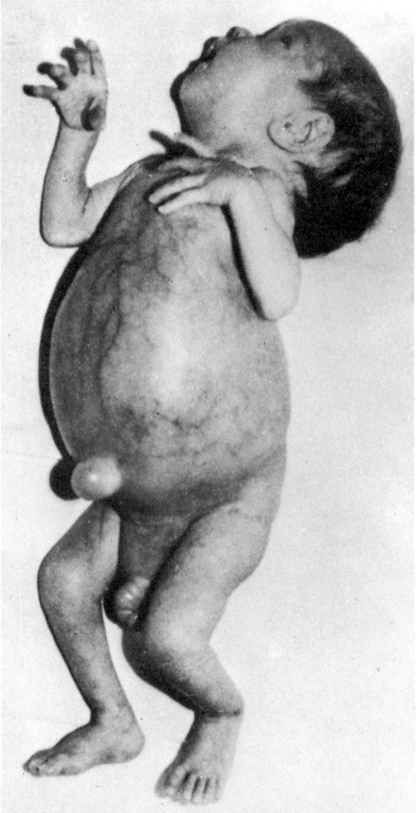

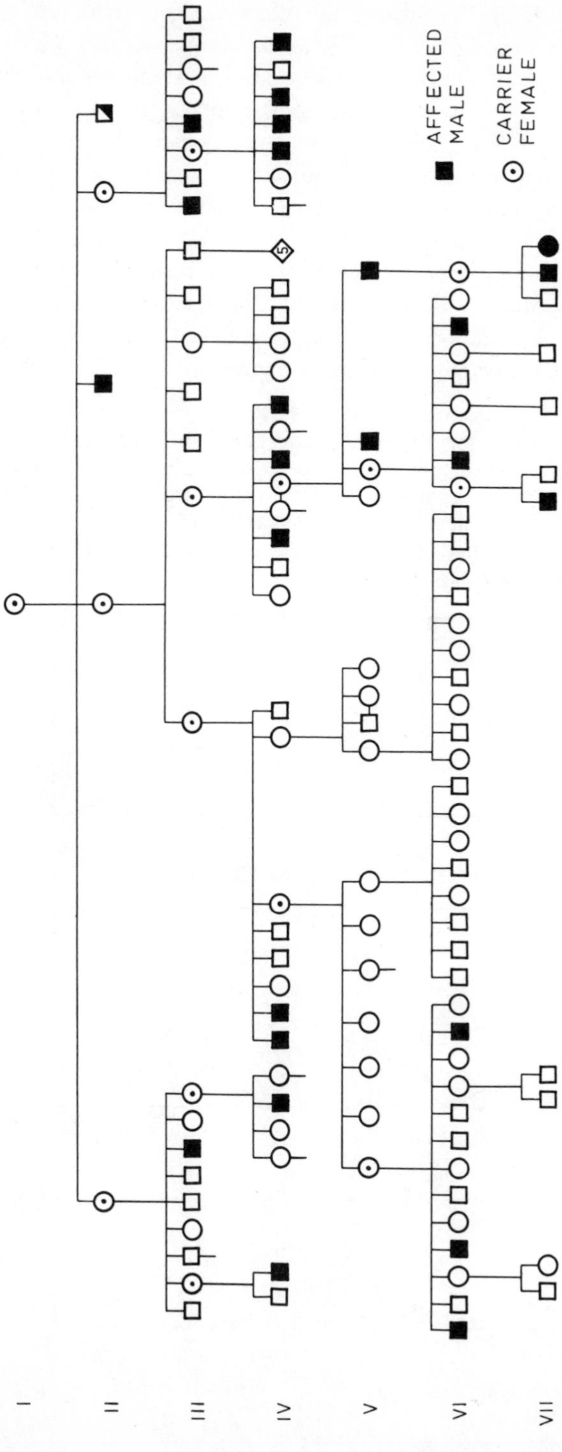

Fig. 3.3 Pedigree of the X-linked inheritance of hemophilia A in a family studied through seven generations. (After E. Ikkala. Scand. Clin. Lab. Investig. **45,** Suppl. 12, [1960].)

population being so great that the gene effect cannot be detected at all. In such a case the gene is said to be nonpenetrant. When pedigrees of a specific trait are studied, there may be some generation in which the typical anomaly does not appear at all; this is called a skipped generation. It is questionable, however, whether true non-penetrance exists. The gene marker, for example, an anatomical defect, is usually the end result of a long chain of events, and the initial failure caused by the mutant gene may have been compensated so well by other genetic or epigenetic factors that the result is indistinguishable from the normal condition, although the primary gene product—polypeptide or enzyme—is as much affected as in the cases where the marker characteristic is clearly expressed. A well-known

Fig. 3.4 Distribution curves for three characteristics in phenylketonuric patients and a sample of genetically healthy people. Intelligence and head size show overlapping, but the curves for the concentration of plasma phenylalanine in the two samples are widely separate. (After W. E. Knox. *In* Metabolic Basis of Inherited Disease, J. B. Stambury *et al.* [eds.] p. 258 McGraw-Hill, Co., New York [1966].)

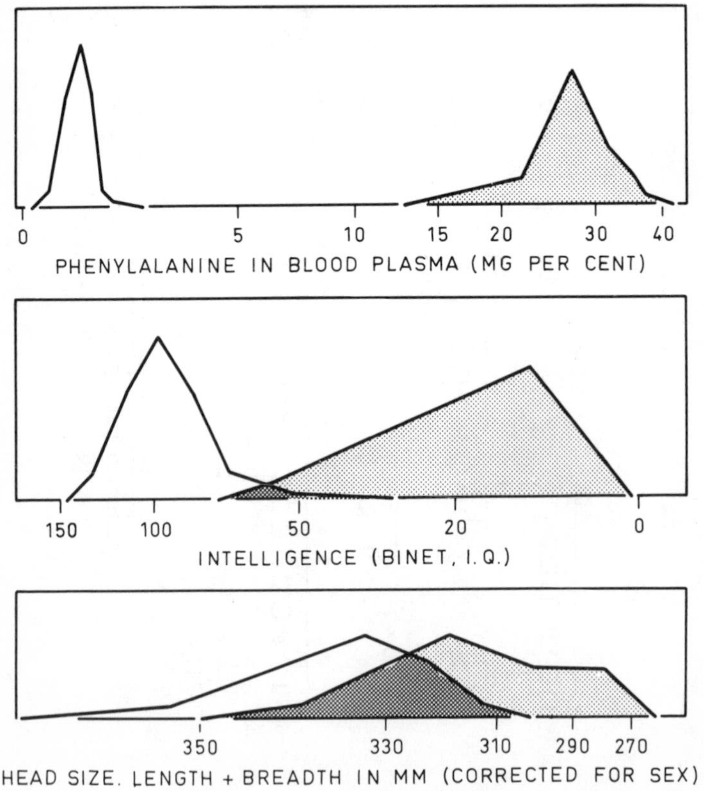

recessively inherited disorder, phenylketonuria, may be taken as an example. Most of the affected individuals are mentally retarded, with a slightly smaller head than the unaffected sibs or the population in general. If mental retardation and head size are used as genetic markers, a considerable overlap is found between the normal population and the phenylketonuric individuals, and the gene can be regarded as nonpenetrant in a fraction of the homozygotes. But if a biochemical marker is used, which brings the testing closer to the primary gene effect, no overlap between the normal and affected populations exists and by this criterion the gene is fully penetrant (Fig. 3.4).

Inborn Errors of Metabolism

In the early years of this century, Sir Archibald Garrod brought together three previously very loosely connected branches of science, namely genetics, biochemistry and clinical medicine, by extensively describing a group of disorders which he called inborn errors of metabolism. His views, greatly in advance of his time, came very close to the fundamental contemporary "one-gene-one-enzyme" or "one-cistron-one-polypeptide" concept of biochemical genetics. Discoveries of several inherited metabolic defects and their etiology at the enzymatic level have largely bridged the gap between the Mendelian principles of inheritance and modern, rapidly advancing, knowledge of the basic molecular mechanisms of gene function.

Inborn errors of metabolism consist of inherited congenital blocks in metabolic pathways traceable to gene-dependent enzymatic defects. One of the main current concepts of cell differentiation, on the other hand, is that this process depends on differential gene activation, first expressed in the synthesis of specific enzymes which ultimately lead to specific biochemical and morphologic characteristics of differentiated cells. With some generalization, all genetic defects could be understood as having a metabolic basis although sometimes limited spatially and temporally to the critical steps of cell differentiation. Direct proof of the existence of such mechanisms is still lacking in higher vertebrates, but the recognized syndromes of metabolic and morphologic defects imply the general existence of such mechanisms.

By now more than 100 different disorders in man have been shown to result from the absence or probable absence of enzymes. The enzymes implicated are concerned with many different types of metabolism, such as carbohydrate, amino acid, lipid, and hormone synthetic pathways. A comprehensive treatment of the metabolic aspects of these disorders will be found in recent textbooks, for example, the one edited by Stanbury *et al.* (1966).

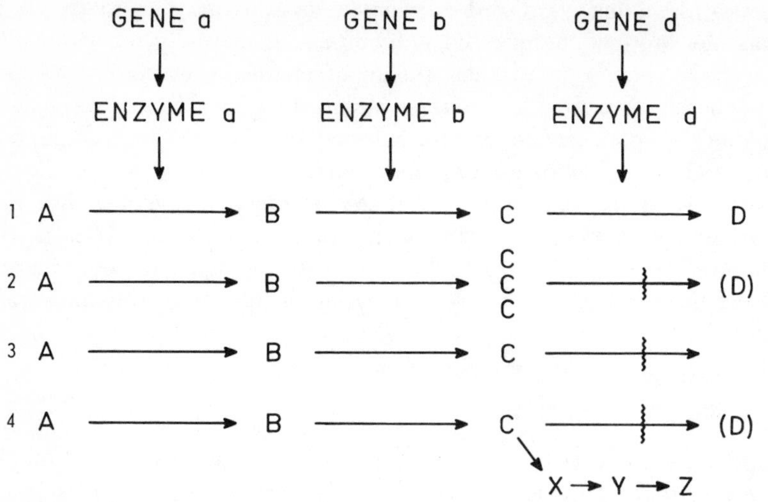

Fig. 3.5 Schematic presentation of the effect of an enzyme block on a hypothetic metabolic pathway. In 1 is shown the normal situation. In 2 the metabolite proximal to the block accumulates in excess. In 3 the metabolite distal to the block is lacking. In 4 the metabolite proximal to the block takes an alternative pathway, normally nonexistent or of minor importance.

The biochemical consequences of enzyme defects are to be seen in the affected metabolic pathways as illustrated in Fig. 3.5. The manifestation and diagnosis of the clinical disorder is usually based on their deficiency beyond it. Two examples of enzyme deficiencies in the metabolism of the amino acid phenylalanine are schematically depicted in Fig. 3.6.

In phenylketonuria the enzyme catalyzing the conversion of

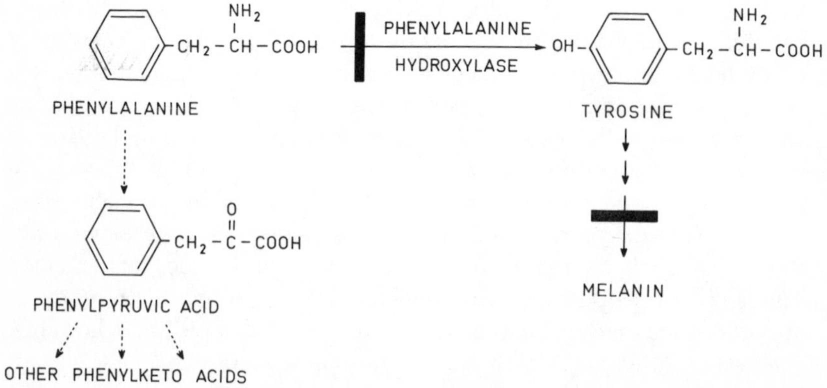

Fig. 3.6 Schematic representation of the enzymatic blocks in phenylalanine metabolism causing phenylketonuria and albinism.

phenylalanine to tyrosine in the liver is lacking and as a result, phenylalanine accumulates in the organism and is then converted through other pathways to abnormal metabolites, which may be found in the tissues, body fluids, and urine. Phenylketonuria is thus an example of the combination of cases 2 and 4 in Fig. 3.5. The phenylketonuric child usually appears normal at birth, but within the first half year shows signs of impaired intellectual development and is later, with few exceptions, severely mentally retarded. The biochemical diagnosis of this defect is based on the large amounts of phenylpyruvic acid excreted in the urine. The chemical basis of the mental retardation appears to be that the abnormal phenylalanine metabolites, as they accumulate, interfere with the metabolism of other aromatic amino acids such as tryptophan. In the other example, albinism, depicted in Fig. 3.6, the disorder manifests itself in the lack of the product of enzyme activity, which leads to a deficiency of melanin production. It is also possible that in some instances the accumulation of the intermediate metabolites, increase in the amount of the substance formed by the alternative minor pathway, and absence of the normal end product all contribute to the pathogenesis of the disease. In the congenital virilizing adrenogenital syndrome (see page 152 and Fig. 6.4) a block in the normal pathway of cortisol synthesis causes accumulation of steroid hormone metabolites, which then, through an alternative pathway, are converted into androgenic hormones. The virilizing effect on the genitalia is due to the excess of androgens. On the other hand, insufficient production of the end product, cortisol, forces the hypophysis to continuous overproduction of ACTH and thus a vicious circle is set up. In some cases it can be interrupted simply by administration of exogenous cortisol.

As previously mentioned, most inborn errors of metabolism are inherited recessively and the heterozygotes do not manifest the disorder. Careful studies have shown, however, that the metabolism of the heterozygous carriers is not entirely normal. Their genome comprises one normal gene and one defective allele incapable of producing the normal enzyme protein. In cases where the enzyme responsible for the disorder is recognizable and its activity can be measured, the heterozygotes have proved to have lower activity than in cases not carrying the mutant gene. Another way to test the metabolic capacity of the heterozygotes is to use a metabolic loading or tolerance test. In such a test the metabolite proximal to the affected step in the pathway is given and its disappearance from the serum or some other body fluid is followed. In heterozygotes the clearance of the metabolite is usually retarded, and it is found in higher concentrations in these subjects than in the healthy controls (Fig. 3.7).

In some inborn errors of metabolism the heterozygote does not have exactly half of the enzyme activity found in unaffected people.

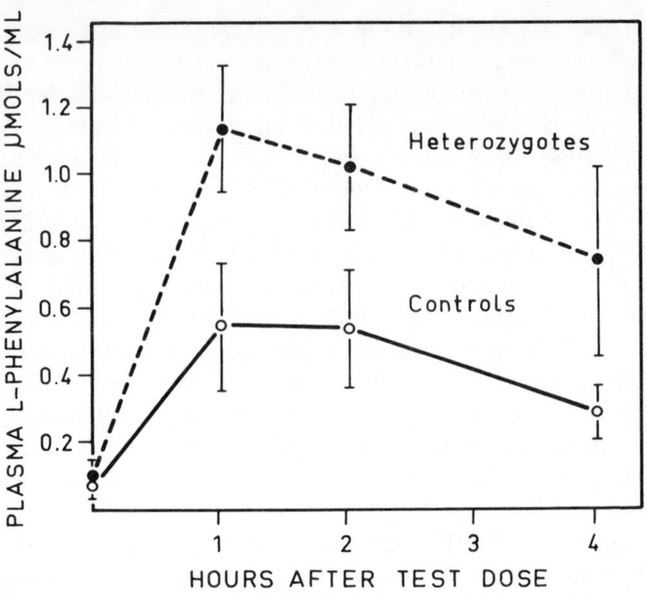

Fig. 3.7 Plasma phenylalanine clearance test in subjects heterozygous for phenylketonuria and in controls. (After W. E. Knox. *In* Metabolic Basis of Inherited Diseases, J. B. Stanbury *et al.* (eds.) p. 258, McGraw-Hill Co., New York [1966].)

Even in the affected homozygous persons considerable amounts of the enzyme may be present, although not sufficient to meet the metabolic needs of the organism. Some propositions, still quite speculative, may be offered to explain these anomalies.

From microbial genetics it is known that the structural gene responsible for the production of enzyme polypeptide does not work alone, but is under the control of regulative genes, as depicted in Fig. 3.8. The basic unit in this scheme is the so-called "operon," consisting of an operator gene and one or more structural genes. The function of the structural gene, RNA transcription, occurs only under the control of the operator. The operator, in turn, is controlled by a regulator gene, which produces a repressor substance (a protein molecule), and this action of the operator is under the control of diffusible cytoplasmic factors. This scheme is based on the pioneering work of Jacob and Monod (1961) and forms part of the rapid advance in modern molecular biology repeatedly reviewed in recent years, (for example, Wagner and Mitchel, 1964; Watson, 1965). From this scheme it is evident that mutations in the regulative genes, as well as in the structural ones, may cause defects in enzyme synthesis. Therefore, the reduced amount of enzyme in an affected person may be due to synthesis of altered enzyme protein with some catalytic

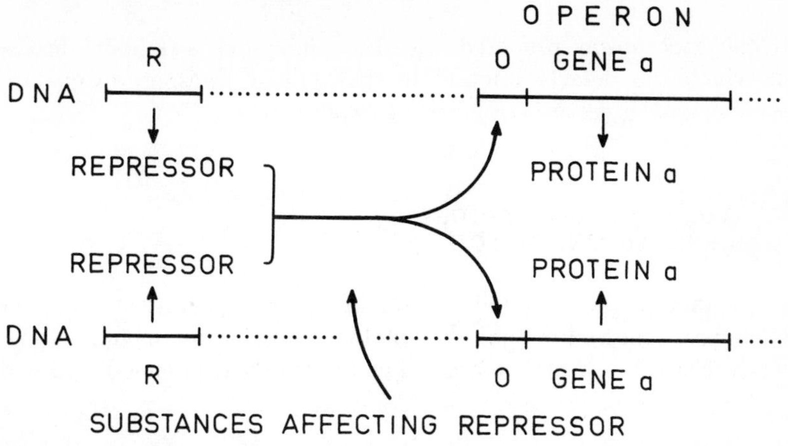

Fig. 3.8 Schematic picture of the functional gene in the diploid organism. (R) regulator gene, (O) operator gene, (a) structural gene.

activity as a result of a mutant structural gene, but it may also be due to abnormal function of regulative genes in the presence of a normal structural gene.

The usually recessive inheritance of inborn errors of metabolism is relatively easy to understand on the one-gene, structural or regulative, one-enzyme theory. The same is not true of the rarer dominantly inherited metabolic diseases. In these cases, heterozygotes show the full effect of the disorder and it is difficult to relate this to lack of some specific enzyme, since the normal allele at least is producing the normal enzyme. In homozygotes absence of the enzyme is possible, but such circumstances generally are not found to exist and it is likely that homozygotes for most of the autosomal metabolic disorders with dominant inheritance are inviable. The nature of the molecular mechanism governing the expression of dominantly inherited metabolic diseases may be suggested on the basis of the operon theory (see Milne, 1966). One possibility is that the affected dominant gene produces the normal enzyme in excess, thus causing accumulation of likewise normal metabolic products in such amounts that they may be harmful to the metabolic economy of the organism. This is the mechanism suggested to account for the dominantly inherited porphyrias. In this case, it is likely that the genetic defect is not in the structural gene, but in the regulative genes. Another possibility is that the mutant gene codes for the synthesis of abnormal polypeptides, which, as structural parts of the cells such as cell membranes or organelles, cause impaired function of the cells themselves, and that the metabolic disorder is secondary to the impaired cell structure. In these cases the defect is likely to be in the structural genes. This is the mechanism proposed for those inherited metabolic diseases in which the defect is found in the cellular transport mechanism, such as renal glycosuria or hereditary spherocytosis.

MULTIFACTORIAL ETIOLOGY OF HEREDITARY DEFECTS

Although the Mendelian laws of inheritance are clearcut and have been shown to apply to all living organisms, including man, many characteristics that are evidently inherited cannot be directly

Fig. 3.9 Incidence of some selected congenital disorders by race in Philadelphia 1961–1962. (From T. H. Ingalls and M. A. Klingberg. Amer. J. Publ. Health **55:** 200 [1965].)

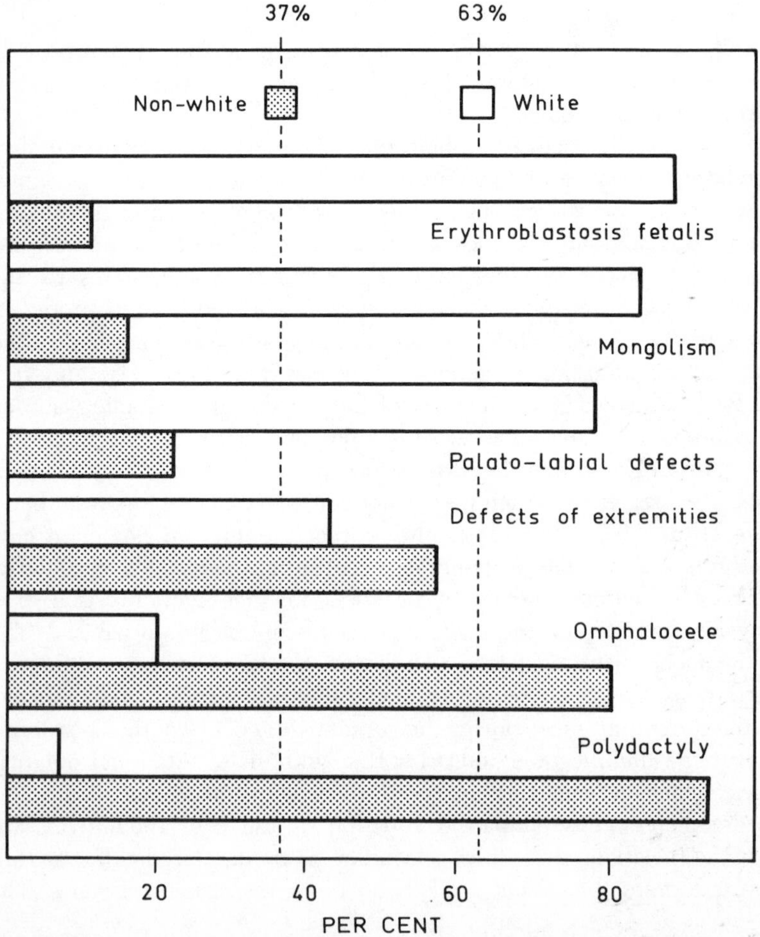

reconciled with them. Still more difficult is the attempt to apply our fragmentary knowledge of molecular gene function to account for the many complex examples of inheritance.

When large series of congenital defects are analyzed, it has been found that many defects have a tendency to recur within the sibships in which they have been found. This recurrence risk is defect-specific, in other words, there is no tendency for increased malformations in general, but a predisposition to the same defect (Neel, 1961). Two examples from human medicine will clarify this situation.

When the incidence of some congenital disorders has been analyzed in large populations composed of different races, a definite differential susceptibility has been found (Fig. 3.9). It seems that certain disorders are manifested more often in the white section of the population (for example, mongolism) while others are more frequent in the nonwhite section (for example, polydactyly). This could be explained on the basis of more frequent dispersion of the genes for specific defects in the respective groups. This may be true but other alternatives, discussed later, also merit consideration.

Another example of a congenital defect in which genetic factors are implicated, but do not follow the simple laws of inheritance, is cleft palate. Cleft palate occurs in white populations in about 1:2500 of all births. The frequency is higher in Japanese and lower in Negro populations. There is a clearcut tendency for recurrence in the affected families, which can be expressed as the risk of the birth of an affected child in the family (Table 3.2).

It is evident that there is a tendency for cleft palate to occur more often in a family with a previously affected member, but this cannot be fitted into the framework of the simple laws of inheritance. The evidence from animal studies, particularly from those performed with inbred strains of mice, sheds some light on this complex situation of inheritance and the possible interplay of inherited and exogenous factors in the etiology of congenital defects.

Gene Interaction

Inbred strains of mice are very useful for genetic studies, because every mouse in the strain is homozygous for each gene pair, except for any new mutations, and thus a gene with a good phenotype marker can be introduced by crosses and backcrosses into a new genetic milieu. Such selection studies have shown that the expression of a single gene is not necessarily an all-or-none phenomenon but depends on interactions with the rest of the genome and that the phenotypes of a known gene may be quite different in different genetic backgrounds.

Genes which modify the manifestation of other genes without

TABLE **3.2**
Risks of Cleft Palate in Various Family Situations[a]

	Risk of Cleft Palate (percent)
Frequency of cleft palate in the general population	0.04
Probability that the next child in the family will be affected if	
both parents are unaffected and	
they have an affected child but no affected relatives	2.0
they have an affected child and an affected relative	7.0
they have an affected child and are relatives	2.0
either of the parents has cleft palate and	
they have no affected children	6.0
they have an affected child	15.0
Risk of some other type of anomaly in any of the above combinations	Same as in the general population

[a] From F. C. Fraser, in Birth Defects, p. 235, (M. Fishbein, ed.), Lippincott, Philadelphia 1963.

themselves having any obvious effect on the normal phenotype are called "modifying genes" or "specific modifiers" (see Grüneberg, 1963). The importance of this type of genetic background is illustrated by the following experiment. The mutant gene *Sd* is expressed in the mouse as taillessness or shortness of the tail and in its original genetic background the gene expression showed great variations. Only a relatively small proportion of the offspring carrying this gene showed total lack of a tail. Dunn (1942) crossed this mutant with normal BALB/c mice and made subsequent backcrosses between *Sd*/+ mice of the F_1 generation and the BALB/c strain. In repeated backcrosses the *Sd* gene was gradually transferred to a new genetic background, which clearly favored its expression, as is shown in Fig. 3.10. The percentage of tailless mice increased steadily in the backcrosses carrying the *Sd* gene, whereas the tails of their normal littermates not carrying the mutant gene were unaffected.

Several other experiments succinctly reviewed by Grüneberg (1963) have elucidated the problem of gene interaction in respect to several genes, and thus seem to be a rather general phenomenon in gene manifestation. The following diagram (Fig. 3.11) gives a concise presentation of the effects of different genetic backgrounds of the mutant gene *a* giving rise to the specific phenotype. In the

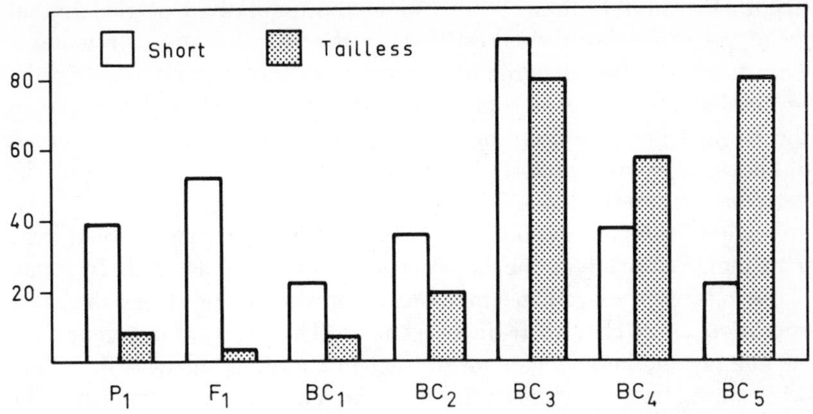

Fig. 3.10 The expression of the mutant *Sd* gene in the original population (P_1) and after transfer to the BALB/c strain in five consecutive backcrosses. (After H. Gruneberg, The Pathology of Development, Blackwell, Oxford [1963], and the original data of L. C. Dunn. Am. Naturalist **76:** 552 [1942].)

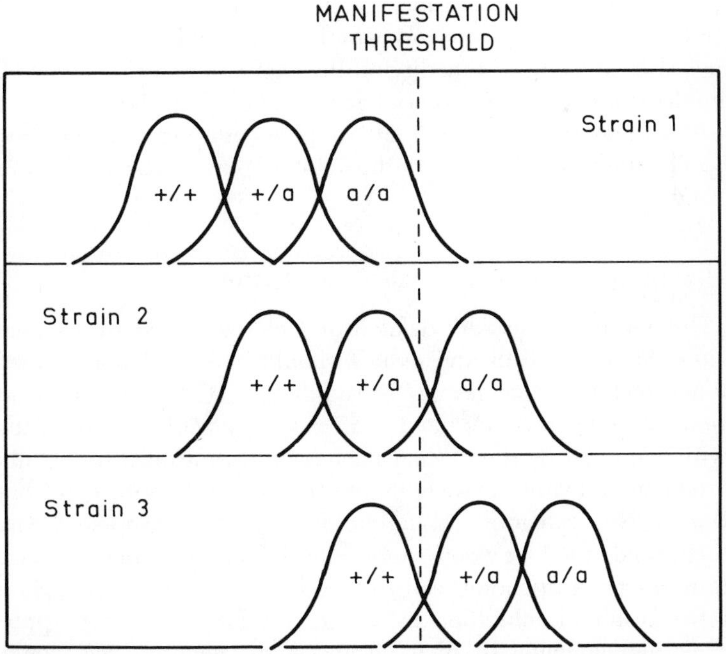

Fig. 3.11 Schematic presentation of the influence of genetic background on the expression of a hypothetic gene *a*. (After H. Gruneberg, The Pathology of Development, Blackwell, Oxford [1963].)

hypothetic inbred strain 1 only some representatives of the mutant genotype *a/a* cross the threshold for the manifestation of gene *a*. The genes of the background genotype in this instance modify the expression of gene *a* in such a way that the effect is detectable in the homozygous condition only. In strain 2 practically all the animals of the *a/a* group manifest the effect of this gene as do some of the heterozygous individuals. The genes of the background in this case are called "dominance modifiers" because they have now enabled gene *a* to manifest itself in the heterozygous condition as well. In strain 3 the background genome in some cases already produces the same condition as is characteristic of gene *a*. The background genes are no longer modifiers of gene *a*, but act in coordination with it. Figure 3.11 shows that the genetic background enhances or diminishes the effect of a particular gene and an intermediate situation or null effect is possible between the two extremes. These results, obtained from studies with inbred strains of mice, may afford some insight into the genetic mechanisms underlying the different frequencies of certain malformations in different races and ethnic groups in man.

A great deal of this book is concerned with the extrinsic factors leading to congenital defects. It may be asked whether, in point of fact, it is legitimate to consider genetic factors and extrinsic factors as acting essentially alone and whether it would not be preferable to regard each as always modifying the expression of the other. There is ample evidence from experimental studies that the latter is true and sometimes the end results of complementary experiments resemble each other so closely that the etiology cannot be guessed from the phenotype alone.

Interplay of Genetic and Exogenous Factors

Differentiation, usually defined and characterized by changes in the morphology of cells and tissues, should be considered as a reflection of preceding changes in the synthetic activity of cells leading to new or enhanced production of macromolecules. This synthetic activity is guided by the cellular genome and controlled by the microenvironment provided by neighboring tissues as described in Chapter 4. Hence, two categories of failure in normal development can be distinguished: the first a consequence of defective or abnormal genetic material in the responding cells or the controlling tissue and the second the result of inhibition by exogenous factors of the expression of a normal genome. In both instances, the consequence would be failure of the synthetic activity of one or several genes and loss of their product. Therefore, it hardly would be surprising to find identical defects, on both metabolic and morphologic levels, caused by genetic defects and by exogenous factors. Such environmentally induced imitations, known as phenocopies, have been repeatedly shown to occur and certain features of this phenomenon will be discussed below.

Another question raised by the above comments would be the possible interplay of genetic and exogenous factors in teratogenesis. If an identical end-result could be obtained from either of the two acting independently, would it not be expected that a defective genome and a phenocopy-inducing teratogen would act synergistically and that a genetically handicapped cell would be more susceptible to the environmental factor than a normal one? We have already presented evidence showing that certain inbred strains of mice differ in their susceptibility to teratogenic treatment (page 20) and when the genesis of malformations is discussed in the next chapter, some hypotheses to account for such interaction of genetic and exogenous factors will be presented (page 82). It is our intention here to quote some recent results concerning such "silent" actions of genes and discuss their possible mechanisms.

The effect of genetic background on the action of environmental teratogens has been demonstrated in a number of experiments, one of which will be reported here. Waddington (1967) has described a series of experiments on Drosophila in which the venation of the wing was used as a marker. Short-term heat treatment during certain sensitive periods of development resulted in alteration of this venation in a rather specific way. However, the extent to which this effect was manifest was not uniform in the original stock and in addition to affected individuals, totally normal ones were frequently found. Subsequently, selection experiments were made in which only flies with the defect were used for further breeding. As a result, a highly susceptible strain was developed, showing a high manifestation percentage of the heat-induced defect. Correspondingly, selection of unaffected individuals for breeding resulted in the development of a resistant strain. Thus, the results show that the reaction to the teratogen is under genetic control and may be influenced by selection like any other genetic characteristic.

Synergistic action of genes and exogenous teratogens A few detailed analyses have been performed on the synergistic action of a mutant gene and an exogenous teratogen. The following two series of experiments have been selected to represent them.

In mice of strain C57BL/10, two mutants, called "luxoid"(*lu*) and "luxate" (*lx*), are known to show a variety of defects, some of them common to both. One of these common defects is tibial hemimelia, which is manifest in homozygous individuals of both mutants.

Anatomically, however, the hemimelia is slightly different and, in addition, the two mutants differ in several other skeletal defects, some of which are indicated in Table 3.3. Dagg (1967), in his experiments, employed mice of these mutant strains (and normal wild-type mice) and tested the effect of a potent teratogen, 5-fluorouracil (FU) (see page 126). In normal C57BL/10 mice a single dose of 500 μg 5-FU given on day 10 of pregnancy produced a malformation syndrome consisting of tibial hemimelia and cleft palate in the offspring. The interesting finding was that a lower dose of 250 μg of the teratogen, which was without effect in normal mice, produced malformations in the heterozygous luxoid and luxate animals (*lu*/+ and *lx*/+). Among the skeletal defects induced were tibial hemimelia and minor defects of the sesamoid bones (Table 3.3), but the cleft palate seen in the series of nonmutant mice was not induced in the *lu*/+ and *lx*/+ heterozygotes.

The main conclusion to be drawn from these experiments seems to be that a teratogen can act synergistically with a mutant gene and induce homologous defects in the heterozygotes in doses which are ineffective in wild type embryos and which are incapable of producing nonhomologous defects in the heterozygotes. The fact that the defect induced by 5-FU was always of the "luxate" type irrespective of which mutant gene was present in the heterozygote remains to be analyzed, as well as does the metabolic mechanism of this phenomenon. Some suggestions will be put forward later in this chapter.

TABLE 3.3
Certain Defects in Untreated Luxoid and Luxate Homozygotes, Fluorouracil-Treated Heterozygotes of the Same Mutants, and Wild-Type Embryos.[a]

Genotype	Treatment	Hemimelia Type A	Hemimelia Type B	Patella Missing	Fabella Missing	Cleft Palate
1u/1u	None	+	−	+	−	−
1x/1x	None	−	+	−	+	−
1u/+	250/μg FU	−	+	−	+	−
1x/+	250/μg FU	−	+	−	+	−
+/+	250/μg FU	−	−	−	−	−
+/+	500/μg FU	+				+

[a] Modified from C. P. Dagg: J. Exp. Zool. 164:479, 1967.

In another series of experiments, Dagg and his collaborators (1966) examined the polygenicity of the genetic control of suscepti-

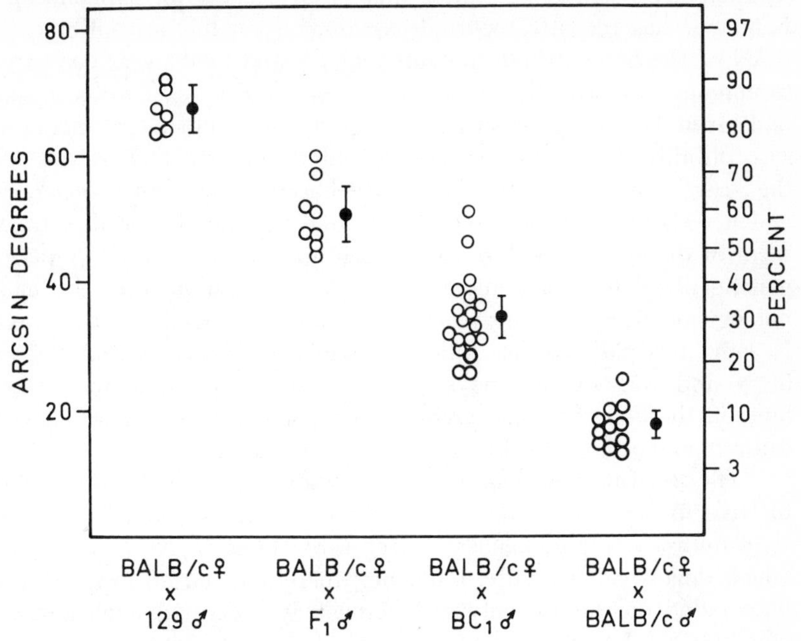

Fig. 3.12 The incidence of polydactyly (as arc sin degrees and percentage) in the F_1 generation of a cross between strains 129/Dg and BALB/c mice, in two subsequent backcrosses to BALB/c and in the BALB/c strain. (After C. P. Dagg *et al.* Genetics **53:** 1101 [1966].)

bility to exogenous teratogens. Polydactyly and cleft palate were induced with 5-FU in two inbred strains of mice, 129/Dg and BALB/c, and the results showed that the former was definitely more susceptible to the treatment. Even by the F_1 generation of a cross between these two strains a relative resistance toward 5-FU had developed as compared to the 129/Dg mice, and two subsequent backcrosses to BALB/c showed a further increase of resistance, as illustrated in Fig. 3.12. The result, moreover, made it possible to calculate the minimum number of loci determining the development of polydactyly and cleft palate. The calculations indicated a minimum of four loci for the polydactyly and three for cleft palate.

As another example, the investigations by Landauer (1965a, 1965b) on mutant chicks should be described. Three different mutant strains are known in which a recessive gene leads to micromelia in homozygous embryos: Californian micromelia (mm^A), Massachusetts micromelia (mm^H) and Lamoreux-type chondrodystrophy (*ch*). The type of micromelia seen in the two first-mentioned mutants greatly resembles that caused by treatment with 6-aminonicotinamide (6-AN);

the defect caused by the latter thus is an example of a phenocopy. Landauer has recently reported experiments on the susceptibility to 6-AN of the heterozygote mutants $mm^{A}/+$ and $mm^{H}/+$ as compared to nonmutant chicks. The mm^{A} mutant behaved much like the luxoid and luxate heterozygotes in Dagg's experiments and showed increased susceptibility during the early stages of development. This was not the case, however, with the Massachusetts micromelia, mm^{H}, in which there was no increased susceptibility to this defect after treatment of the embryo with 6-AN. Thus it appears that the two types of geneticially determined micromelia differ in their genesis, the morphologic end-result being reached by different metabolic pathways. In mm^{H} this pathway has little in common with the one induced by 6-AN and consequently no synergism occurs. To quote the author himself, the gene and the teratogen "do not share steps at which one can promote or impede effects caused by the other."

The third mutant gene mentioned above (*ch*) has been reported to have incomplete expressivity in the original stock and in addition to chondrodystrophic chicks, there were some *ch*/*ch* homozygotes which developed normally. Landauer, therefore, started to select two lines based on this phenotypic heterogeneity, one with high and another with relatively low expressivity. These two lines were subsequently tested for their susceptibility to 6-AN as above and some of the results are indicated in Table 3.4. The results show that selection against extreme expressivity of the gene *ch* has led to a line with greatly reduced susceptibility to a teratogen causing homologous defects. The author concludes that the modifying genes which suppress the homozygous expression of *ch*/*ch* also reduce the susceptibility to 6-AN. The example clearly shows how the silent action of a single gene, expressed as increased susceptibility to exogenous teratogens, should always be considered in relation to the genetic background as a whole.

The results quoted above suggest that, in cases where synergistic action between a mutant gene and an exogenous teratogen can be demonstrated, the biochemical pathways by which the maldevelopment is induced by the two must be sufficiently closely related to permit interaction. Little is known about these metabolic pathways in most instances, but brief speculation on the subject is permissible.

As stated on page 50, differentiation at the morphological level merely represents the synthesis of new macromolecules or an increase in their production. As discussed elsewhere in this book, it is conceivable that such increasing synthesis during embryonic development must reach a certain threshold level to become expressed as overt differentiation. Below this level we may still have a specific synthetic activity without manifest differentiation. At the beginning of this

TABLE 3.4
The Susceptibility of Two Mutant Lines Obtained through Selection against High and Low Expressivity of the Chondrodystrophic Gene *ch*.
Heterozygotes were treated with 6-aminonicotinamide at different stages of development.[a]

	Treatment at Hours of Incubation		
	84	96	120
	(Percent of Micromelic Embryos)		
Low expressivity line	10	40	13
High expressivity line	37	65	13

[a] From W. Landauer: J. Exp. Zool. **160**:345, 1965.

chapter we noted that both an exogenous factor and a defect in the genome that controls such specific metabolic activities can inhibit the production of macromolecular compounds and consequently induce overt malformations. In our examples of recessive mutants, it appeared that the synthetic activity controlled by the nonmutant gene is sufficient to bring production of the corresponding protein(s) above the threshold, whereas the homozygote genome failed to do so. It may be of interest here to note that in certain conditions caused by autosomal mutations in man, where the gene products can be quantitated, such lowered synthetic activities can be directly measured. In these conditions, the heterozygote appears to produce approximately one-half of the quantity of the protein normally synthesized, whereas the homozygote mutant shows only traces of protein. Good examples of such conditions are the hemoglobinopathies and the syndrome caused by lack of thromboplastin antecedent (PTA).

After these comments, we may reconsider the results obtained by Dagg in the two mutant strains of mice and speculate on the way in which 5-fluorouracil affects them (Fig. 3.13). We assume the specific synthesis of a certain product or products required for normal development of the tibial region and of a threshold level which has to be reached during embryogenesis if normal limb morphogenesis is to occur. The gene(s) controlling this activity is lacking or defective in the *lx* and *lu* mutants and consequently a homozygous recessive genome never allows the production of our hypothetic protein to reach the threshold, whereas in heterozygous embryos the decreased synthesis is still high enough to reach the differentiative level. Treatment with the metabolic inhibitor 5-fluorouracil affects this synthesis in both wild-type, nonmutant embryos and heterozygotes but in the latter it acts synergistically with the defective genome, with the result

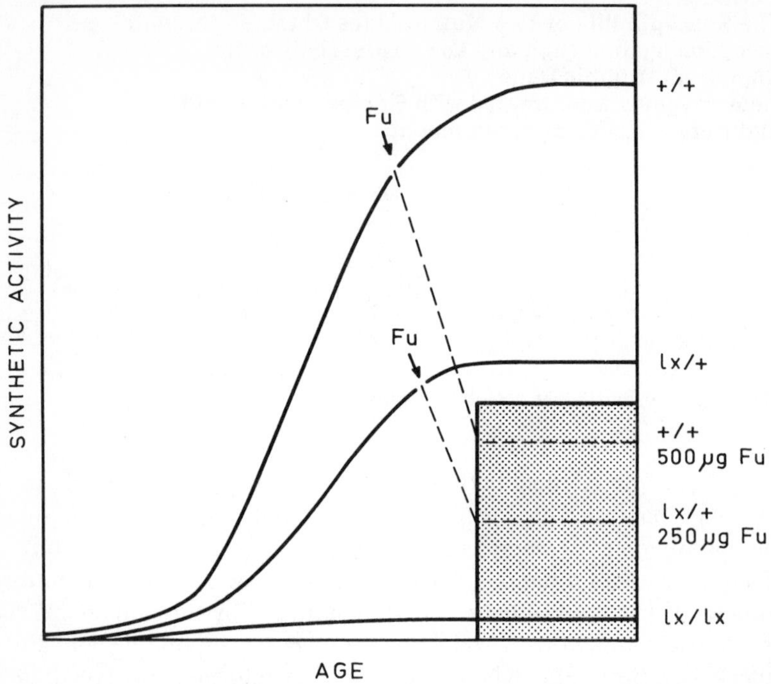

Fig. 3.13 Hypothetic explanation of the results obtained with 5-fluorouracil (FU) treatment in a mutant (*lx*) and a nonmutant (+) strain of mice, described in detail in the text.

that a lower dose is sufficient to prevent the synthetic activity from reaching the threshold value.

This scheme may give an idea of the type of interaction we are looking for, but it is certainly an oversimplification. To analyze such gene-environment interactions new model systems will have to be developed and the possibilities of cell and organ culture methods especially should be explored.

Another possible mechanism of interplay in a specific defect will be discussed in detail in Chapter 4 in connection with the experimental induction of cleft palate.

We may close this section of interactions between genetic and nongenetic factors with two generalizations recently made by Landauer (1965b).

1. An embryo which is heterozygous for a recessive mutation or carries a mutant gene of low penetrance exhibits an increased susceptibility to exogenous teratogens with homologous effects.

2. Modifying genes may induce similar alterations in the expression of a particular mutant gene and of the corresponding phenocopy induced by exogenous factors.

CHROMOSOMAL ABNORMALITIES

Morphologically detectable abnormalities in the chromosomes—both numerical aberrations and structural changes—have attracted increasing interest in medicine and biology for two reasons. First, many congenital disorders have been convincingly attributed to such abnormalities and their mechanism has been clarified to some degree. Second, to a biologist, these abnormalities represent important markers for lesions of early development which have been successfully applied in the analysis of certain "genetic" defects. No doubt, these gross abnormalities represent only the "peak of an iceberg," while most of the defects remain invisible as far as the morphology of the genetic material is concerned but they nevertheless merit the teratologists' full attention.

Frequency of Chromosomal Abnormalities

The overall incidence of chromosomal abnormalities responsible for birth defects in the general population has been estimated to be of the order of 0.5 percent (Polani, 1963). In fact, the incidence of these abnormalities in fertilized eggs is much higher, as the majority of the chromosomal errors lead to nonviable zygotes which are lost during early pregnancy. Spontaneous human abortions, therefore, offer a most interesting opportunity to study such abnormalities and their frequency during early embryogenesis.

Inhorn (1967) has compiled the cases published from 1961 through 1965 and found altogether 466 cytologically analyzed, spontaneously aborted human embryos. In these, chromosomal abnormalities were reported in 120 cases and the author concludes that one-fourth of these abortions could be attributed to chromosomal errors. More recently Carr (1967) has presented his own material based on 227 spontaneous abortions and reports chromosome anomalies in 22 percent of these cases. If the total incidence of abortions in the general population is estimated as 20 percent (Chapter 1), the above studies would add another 5 percent to the overall incidence of chromosomal abnormalities.

As will be seen later, the repertoire of chromosomal abnormalities in abortions and live births is rather limited and this suggests that certain types of gross errors have been eliminated during the early stages of development. On the other hand, we cannot assume that all types of abnormalities originally occur with the same frequency; in fact, a predilection for certain chromosomes and certain parts of chromosomes has been suggested.

Forms of Chromosomal Abnormalities

Let us first briefly present the anatomy of the problem and list the main types of structural and numerical defects in the chromosomes. The list is mainly based on the recent monograph by Bartalos and Baramki (1967), in which the reader will find a more comprehensive description.

Numerical aberrations The total number of chromosomes may vary either as an exact multiple of the basic haploid number (n) of the species (euploidy) or by deviating from this multiple (aneuploidy). The most common situation of the former type is "polyploidy," a state characterized by the presence of more than two haploid sets of chromosomes (3n, 4n, and so on), which may be due to various mechanisms leading to chromosomal duplication without subsequent cell division. Aneuploidy includes both "hypoploid" types where one or several chromosomes has been lost ($2n - a$), and "hyperploid" types, where the nucleus has acquired extra chromosomal material ($2n + a$). The condition in which one member of a pair of chromosomes is missing ($n + n - 1$) is referred to as "monosomy." The chief abnormal mechanisms leading to aneuploidy are those called *nondisjunction* and *anaphase lagging*. In the former, the chromosomes fail to separate during anaphase, with the consequence that one daughter cell receives an extra chromosome, whereas it will be lacking from the other. In the case of anaphase lagging, the chromosomes separate normally, but one is delayed in its migration towards the pole, remains behind the others and is eventually eliminated. In both cases the abnormal mechanism produces two genetically different cell lines, one lacking the unseparated or lagging chromosome and the other with either a normal or a hyperploid chromosome number. If this takes place during early embryogenesis and both cell lines are viable, the result is a genetically inhomogeneous cell population in the organism. Such cases of genetic *mosaicism* are being found in increasing numbers in man, particularly in regard to the sex chromosomes of patients showing various disorders; these defects will be briefly dealt with at the end of this section (Fig. 3.23).

Structural abnormalities Most structural abnormalities of the chromosomes are considered to be consequences of chromatid or chromosome "breaks," which can be experimentally induced by a variety of exogenous factors. The broken ends of the chromatid or chromosome have a strong tendency to "heal" by becoming attached to other similarly broken ends. If union takes place at the normal site, abnormality is not inevitable and if material is exchanged with the homologous pair, the phenomenon corresponds to normal crossing over. If, however, the broken end becomes attached at an abnormal site in the broken chromosome or its homologous pair, or is transferred to a heterologous chromosome, an aberration is created. Figure 3-14 illustrates schematically some such structural abnormalities attributable to chromosomal breakages and their pathologic healing and reattachment. In addition, the formation of *isochromosomes* as a consequence of a misdivision of the centromere is illustrated. The formation of *ring chromosomes* apparently involves two breakages and reunion of the broken ends of the same chromosome.

Causes of Chromosomal Abnormalities

Both nonexperimental, epidemiologic studies and experimental investigations, *in vivo* and especially *in vitro,* have been employed in the search for the causes of chromosomal errors and, in fact, use has been made of many of the methodologic possibilities described in the preceding chapter. A variety of both endogenous and exogenous factors have been suggested and suspected, and some of them have even been proven to play a role in at least experimental conditions.

Epidemiology Most of the epidemiologic studies have been concerned with the most common of all human chromosomal abnormalities, Down's syndrome, which is caused by extra material of the G-group chromosomes (Fig. 3.20). This relatively common disorder so far is the only one for which large series can be collected and the data subsequently related to information on a variety of environmental and socio-medical factors. A significant clustering in time and space of this syndrome has been reported in England and Australia and a nonrandom distribution in season of birth has occasionally been observed (see Day, 1966). On the other hand, negative results indicating random distribution in time and space have also been reported. In attempts to relate the occurrence of this syndrome to other nonrandom factors, both viruses (rubella and hepatitis epidemics) and background radiation have been suggested, but their causative role is far from established. In other reports correlations have been sought

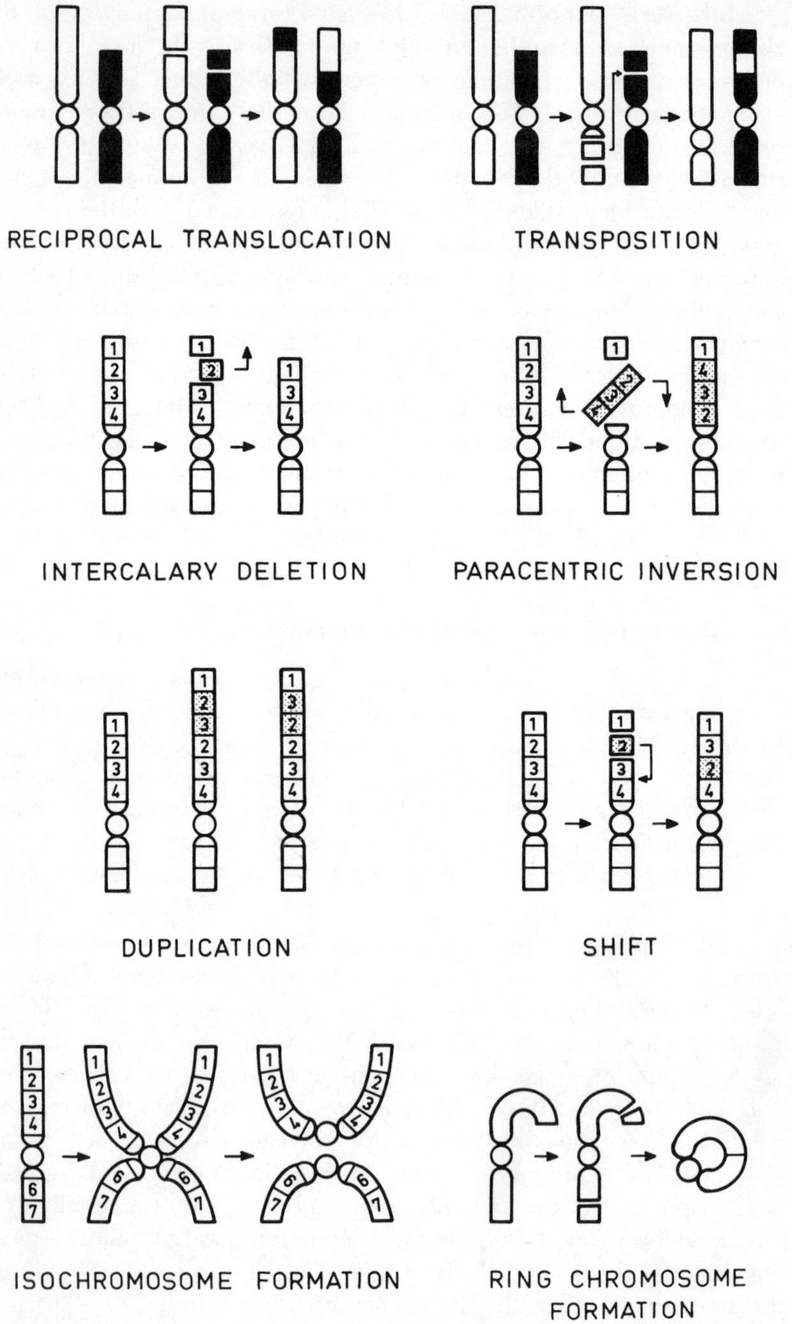

Fig. 3.14 Schematic illustration of some chromosomal abnormalities and their formation.

between chromosomal aberrations and medical factors, social class, and so on, but without consistent results.

So far maternal age seems to be the only factor recognized to be significantly correlated with the birth of children showing Down's syndrome. It has repeatedly been shown that the mean age of mothers giving birth to children with this disorder is definitely higher than that of parturients in general (Fig. 3.15). A similar correlation has more recently been shown between E-trisomy and maternal age. The mechanism behind these correlations is open to speculation, but very recently an interesting hypothesis has been put forward. As will be described in Chapter 5, experimental evidence indicates that delayed fertilization of the oocyte can lead to defective zygotes, nonviable embryos and chromosomal abnormalities. Consequently, as German (1968) suggests, sporadic or decreased frequency of coitus accompanying advanced age may lead to such defects through delayed fertilization.

Radiation In view of the known mutagenic action of ionizing radiation, this exogenous factor might well be one of the first to be suspected of causing chromosomal abnormalities. As *in vivo* studies are still scanty (see Chapter 7), we may take some examples from *in vitro* studies perhaps not directly related to teratogenesis.

Fig. 3.15 The age distribution of parturients in the general population and of those giving birth to children with Down's syndrome. (After V. A. McKusick. Human Genetics Prentice-Hall, Englewood Cliffs, New Jersey [1964].)

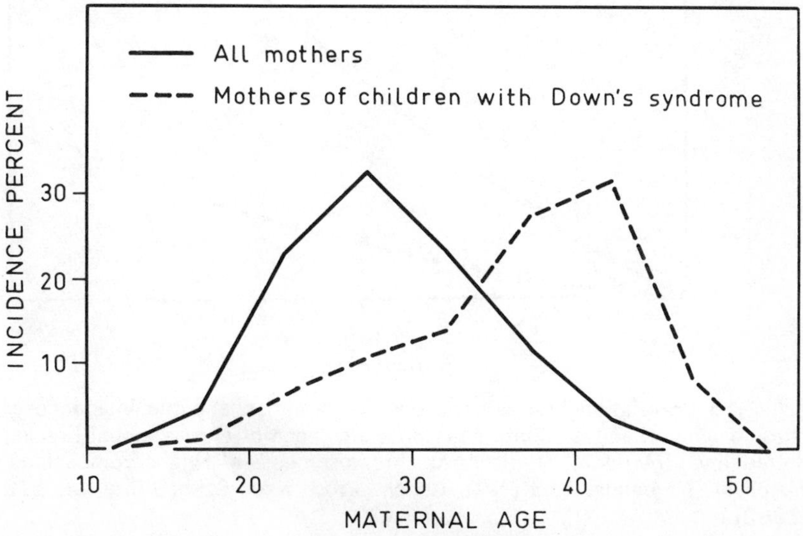

When human peripheral leukocytes were irradiated *in vitro* during the G1 phase of the cell cycle preceding DNA synthesis, chromosomal breaks and aberrations were observed in the subsequent metaphase. The frequency of these increased almost linearly with the dose of radiation (Fig. 3.16). Some of the anomalies were labile and disappeared from the culture rapidly, whereas others were extremely stable and persisted *in vitro* for long periods. In the peripheral blood of human patients exposed to high doses of radiation, such stable abnormalities have been seen to persist for 10 to 15 years and they may even be used as a biologic dosimeter for calculating the original amount of irradiation (Buckton *et al.*, 1962).

Chemical factors *In vitro* studies performed on mammalian cell lines have revealed that a great number of chemical compounds cause chromosomal breaks and structural aberrations. These agents include substances that directly interfere with DNA synthesis (alkylating agents, fluorodeoxyuridine, Mitomycin C and Actinomycin D) and

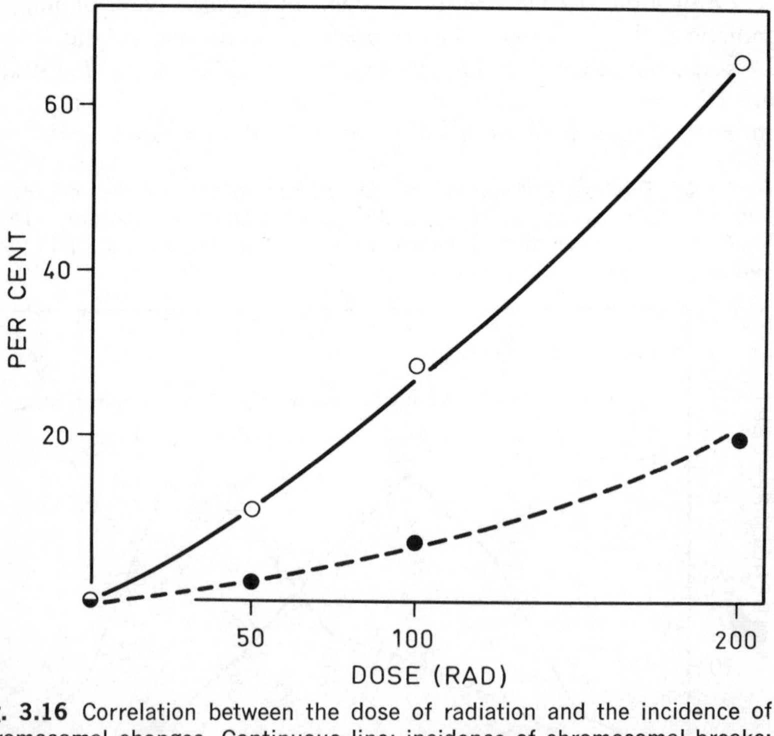

Fig. 3.16 Correlation between the dose of radiation and the incidence of chromosomal changes. Continuous line: incidence of chromosomal breaks; dotted line: frequency of dicentric chromosomes and ring chromosomes. (After M. A. Bender and P. C. Gooch. Proc. Natl. Sci. (U.S.). **48:** 522 [1962].)

are known to be potent teratogens (Chapter 8). Moreover, inhibitors of protein synthesis, for example, certain alkaloids and Puromycin, induce chromosomal abnormalities, as do mitotic inhibitors (colchicine). It is interesting to note that most of these chemical compounds are known to cause mutations (Auerbach, 1967), suggesting that similar effects on DNA may be involved in both visible chromosomal abnormalities and morphologically undetectable point mutations.

Chromosomal and chromatid breakages induced by chemical factors are usually randomly distributed over the chromosomes like those produced by radiation, but we may cite an interesting example of an exception to this rule. In certain chromosomes of the Chinese hamster, the different segments are relatively easy to distinguish and consequently breaks induced in them can be mapped. Figure 3.17 illustrates the results of an experiment in which the effects of three exogenous factors on these chromosomes have been compared: 5-bromodeoxyuridine (BUDR), hydroxylamine (HA), and irradiation. The results clearly indicate a nonrandom distribution of the breaks induced by the two chemicals, whereas the distribution noted after irradiation was random. As BUDR specifically replaces the thymidine in DNA and HA correspondingly interferes with the cytosine, it may be concluded that segment 7, affected by the former, contains abundant adenine-thymidine pairs, whereas segment 5 seems to be rich in cytocine-guanine pairs. This example shows that a rather specific and restricted effect can be obtained with an exogenous factor, suggesting that a correlation between a specific chromosomal aberration (disease) and an exogenous factor is not, in fact, beyond the bounds of possibility.

Viruses Virus-induced chromosomal abnormalities mainly consist of chromatid-type breaks with a notably poor tendency to rejoin. In addition, certain viruses may cause severe destruction, referred to as pulverization, where all or part of the chromosomes become totally fragmented. Examples of these lesions are illustrated in Figs. 3.18 and 3.19. A third group of virus-induced abnormalities include pathologic mitotic figures and abnormal spindle formation following nuclear fusion (see Nichols, 1966).

Not only have chromosomal abnormalities apparently induced by viruses been seen in experimentally infected cells, but similar aberrations have been reported in the cells of the peripheral blood of human subjects who have contracted virus infections or been vaccinated with live virus preparations. Such effects have been seen in connection with the following infections: chickenpox, mumps, measles, and hepatitis (vaccination) (see Aula, 1965).

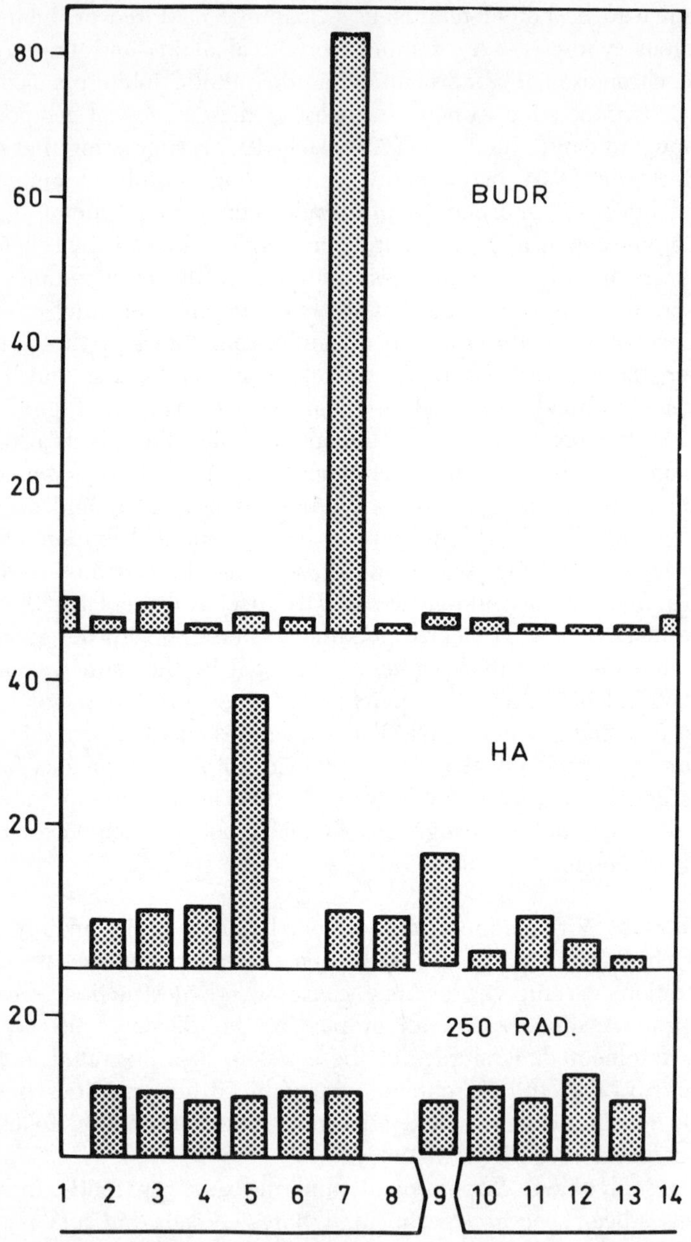

Fig. 3.17 The distribution of chromosomal breaks in chromosome No. 1 of the Chinese hamster after different treatment. BUDR: bromodeoxyuridine; HA: hydroxylamine. (After C. E. Somers and T. C. Hsu. Proc. Nat. Acad. Sci. (Wash.) **48:** 937 [1962].)

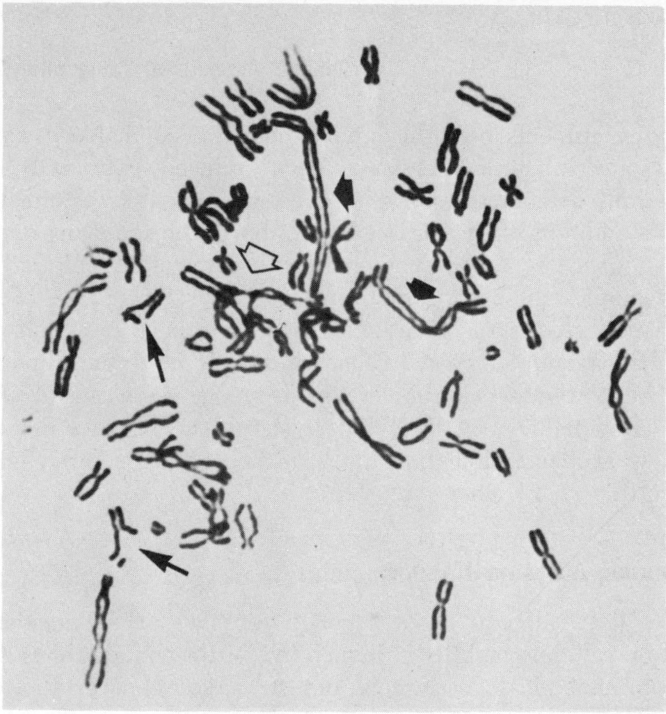

Fig. 3.18 Examples of virus-induced chromosomal aberrations. Small arrows represent chromatid breaks, thick arrows indicate dicentric chromosomes, and the open arrow points to a triradial type of aberration. (Courtesy of Dr. Eero Saksela.)

Fig. 3.19 Pulverization of chromosomes caused by parainfluenza 1 (Sendai) virus in cultured human neoplastic cells. (Courtesy of Dr. Eero Saksela.)

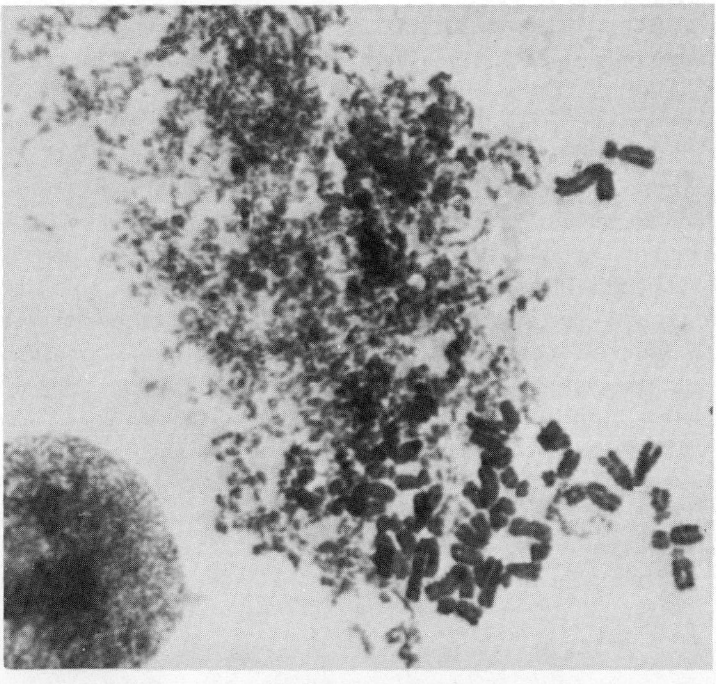

In experiments performed on human and animal cells *in vitro*, chromosomal abnormalities have been induced both with benign viruses and with those known to cause malignant transformation of cells. In addition to the viruses listed above, the following ones may be mentioned as causing chromosomal damage in cultured cells: herpes simplex, rubella (german measles), parainfluenza 1 (Sendai), Newcastle disease virus, mouse and fowl leukosis viruses, Simian virus 40, polyoma and Shope papiloma viruses. It is of some interest to note that the teratogenic rubella virus is among those causing chromosomal abnormalities *in vitro*, but it should be stressed that so far very few similar findings *in vivo* have been reported after infection with rubella (Nusbacher *et al.*, 1967).

Human Autosomal Abnormalities

We have already given a rough estimate of the incidence of chromosomal abnormalities in human live births and abortions, including both numerical aberrations and structural abnormalities. Especially in abortions, a wide variety of abnormalities of different types have been reported, but at later stages of development this impressive array is reduced, as most of the changes are apparently lethal and do not allow the embryo to survive to full term. Consequently, relatively few profound anomalies are known in human beings, although there is a great variety of minor structural abnormalities.

Autosomal trisomy Of the 22 theoretically possible autosomal trisomies, only three have been found in liveborn children, namely trisomies of the D-, E- and G-group chromosomes. Of these, the G trisomy, known as Down's syndrome, is by far the most common and consequently the best known. The syndrome consists of mental retardation, a typical "mongoloid" facies, and a variety of minor, inconstant anomalies; its incidence among Caucasian people has been estimated to be of the order of 1:700. In more than 95 percent of these mongoloids, the karyologic basis of the disorder is a true trisomy of the group G chromosomes (21–22), probably caused by nondisjunction. In some cases, however, the extra G material is found translocated to other chromosomes, usually to those of the D group (Fig. 3.20). The latter form could consequently be expected to be a familiar one, as suggested by the scheme in Fig. 3.21 and known from clinical experience and published pedigrees (for example, Jackson and Ashford, 1967). As shown in this figure, this translocation of G material leads to phenotypically normal carriers of the disease who may carry the extra material in their germ cells. In cases of paternal carriers,

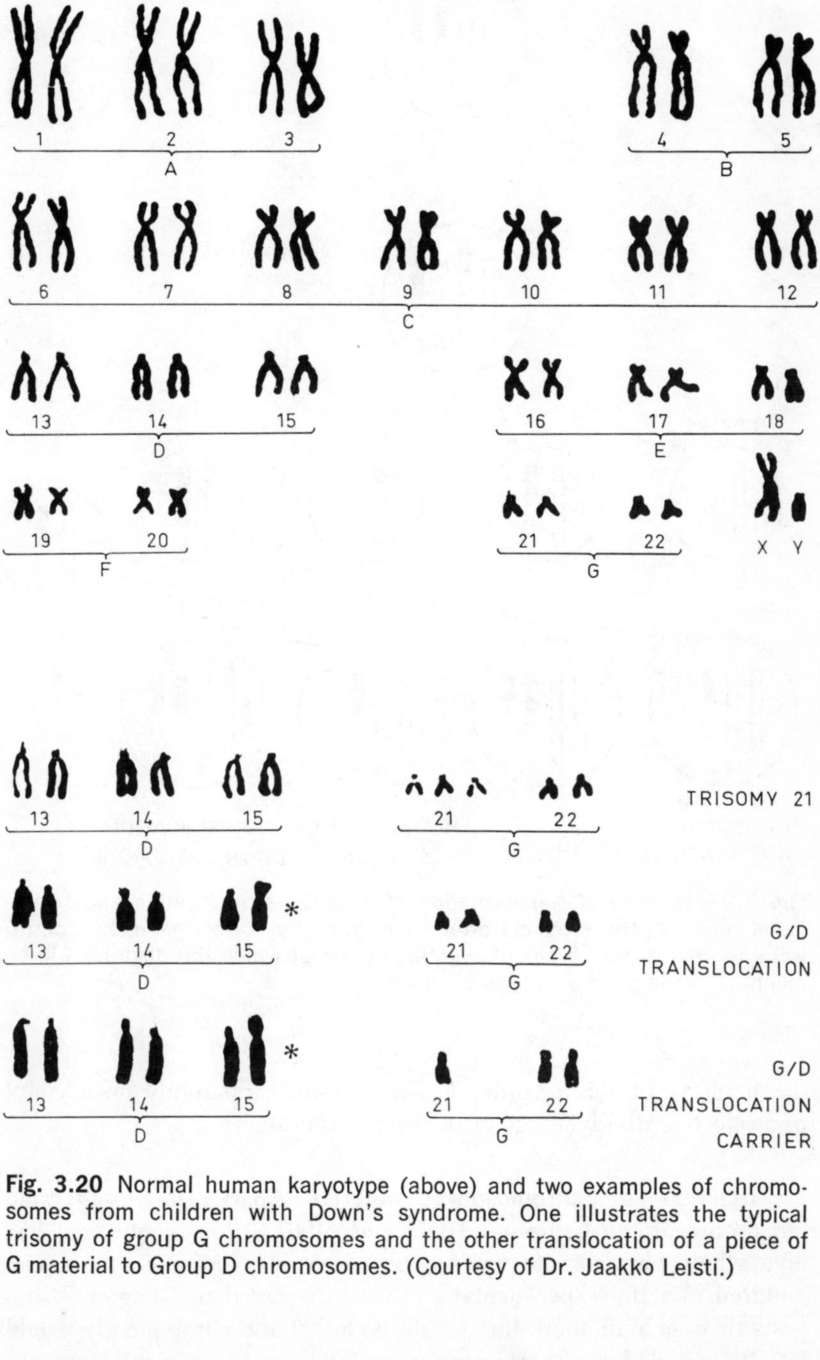

Fig. 3.20 Normal human karyotype (above) and two examples of chromosomes from children with Down's syndrome. One illustrates the typical trisomy of group G chromosomes and the other translocation of a piece of G material to Group D chromosomes. (Courtesy of Dr. Jaakko Leisti.)

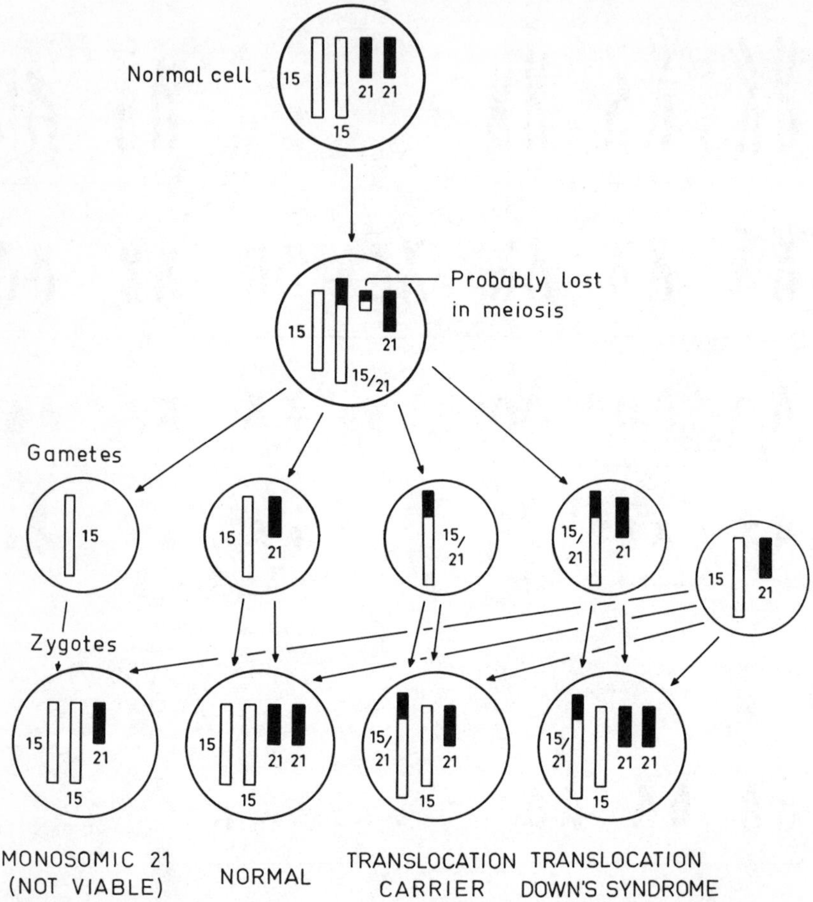

Fig. 3.21 Scheme of the formation of a 15/21 translocation and subsequent meiosis. The products after fertilization by normal gametes (sperm) will give rise to four types of zygotes, in one of which the disorder will be manifest, while another will be a carrier.

the heredity of this disorder is less obvious, apparently on account of a selective disadvantage of the carrier sperm.

Deletion of chromosome material Apart from suggested monosomy of the G-group (Hall *et al.*, 1967), the unequivocal loss of a whole chromosome has not been described in liveborn human children and the experimental evidence discussed in Chapter 7 suggests that such an aberration would be lethal and consequently would be eliminated during early pregnancy. Deletions of minor fragments, on the other hand, have frequently been found in children with serious congenital defects. So far, only one relatively constant syndrome associated with a certain deletion has crystallized from this heterogeneous material. In this so-called "cri du chat" syndrome, the deleted material is missing from the short arm of one of the group B chromosomes (4–5) and this is correlated with a complex defect syndrome with multiple malformations. The most common abnormal feature of these patients is their peculiar cry, greatly resembling a cat mewing, but microcephaly, widely spaced eyes, and certain other defects seem to be relatively common in the syndrome.

Autosomal mosaicism Again, a great variety of cases have been reported showing autosomal mosaics and there is no need to list them here. An interesting feature is the appearance of certain lethal aberrations in the mosaic organism. Several examples of such aberrations are known where the abnormal karyotype does not appear to be viable on its own, but seems only to survive in a mosaic organism. These cases suggest that the normal cell population exerts some kind of sustaining effect on the abnormal cell line, preventing them from succumbing.

Abnormalities of Human Sex Chromosomes

Nearly two-thirds of the known chromosomal anomalies in liveborn children consist of abnormalities of the sex chromosomes, most of which are numerical aberrations. We shall not devote space to an account of the development and clinical symptoms of these different disorders; for such descriptions the reader should consult specific monographs relating to this field (for example, Federman, 1967). Only some general comments will be made.

Sex chromosomes It is generally believed that the small Y chromosome is entirely concerned with the determination of sex and does not carry genetic information other than that involved in male sexual development. The large X chromosome, on the other hand, contains about 5 percent of the total nuclear DNA and carries a variety of genetic messages not directly related to sexual development, as indicated by the more than 75 known hereditary disorders linked to aberrations in the X chromosome. Hence, defects not restricted to sexual development would be expected when the X chromosome is involved in chromosomal aberrations. This is, in fact, the case and in Fig. 3.22 we have listed some of the common sex chromosome abnormalities, together with some general symptoms. Consequently,

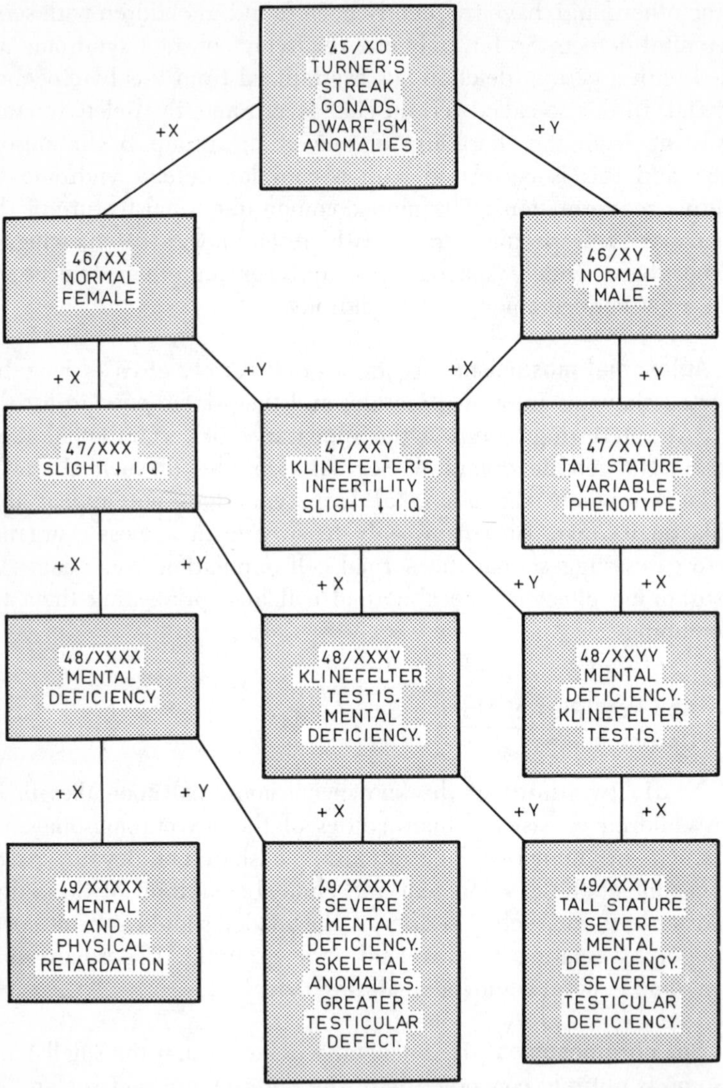

Fig. 3.22 A scheme of sex chromosome anomalies in man as a function of the quality and quantity of chromosomal material. (After D. D. Federman. Abnormal Sexual Development, W. B. Saunders Co., Philadelphia [1967].)

geneticists have wondered why the normal female genotype XX does not phenotypically represent a chromosomal abnormality showing expressions of the extra X material. In recent years much light has been shed on this old paradox and there is increasing evidence to corroborate the "Lyon hypothesis" presented in 1961. According to this, the extra X in human embryos becomes randomly inactivated early in embryogenesis, and results in a certain mosaicism of cells with maternal and paternal X-chromosome expressed. However, inactivation of the second X chromosome seems not to be complete and its genes may be partially expressed, at least during very early development. In addition, the degree of inactivation seems to vary in different parts of the organism. Morphologically, the inactivated material is seen as the typical sex chromatin (Barr body), a small dense nuclear body representing aggregated, nonexpressible material of the extra X chromosome.

It is quite apparent that the Lyon hypothesis can be applied just as well to sex chromosome abnormalities where more than two X chromosomes are present. The fact that the number of Barr bodies in these cases corresponds to the number of extra X chromosomes suggests that all extra X chromosomes are inactivated; and it is of interest to note that if abnormal X chromosomes are present, they are the ones that are preferentially inactivated. On the other hand, the severity of the disorder in these abnormalities seems to be correlated with the number of X chromosomes, as indicated in Fig. 3.22, thus corroborating the idea of a partial inactivation, as mentioned above.

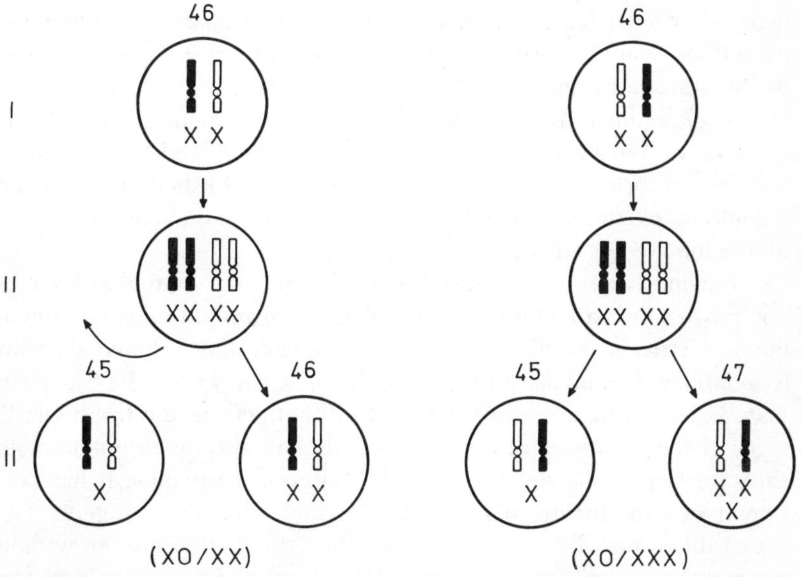

Fig. 3.23 Scheme of the development of XO/XX and XO/XXX mosaicism through anaphase lagging and chromosome nondisjunction. The autosomal complement is indicated as schematic blocks in black and white.

Types of sex chromosome abnormalities Figure 3.22 illustrates several sex chromosome abnormalities in human beings and their clinical features. In Chapter 6 we will briefly return to this question and comment on various endocrinologic aspects of these disorders. In addition to the types illustrated in the figure, a great variety of sex chromosome mosaicisms have been described and Fig. 3.23 schematizes the way in which two such conditions may arise through anaphase lagging and nondisjunction of the chromosomes. As can be expected, the clinical picture of these mosaic cases varies greatly and depends, among other factors, on the degree of mosaicism and the amount of extra chromosomal material.

MATERNAL-FETAL INCOMPATIBILITY

In this chapter it has been shown that the resistance or susceptibility of the fetus to gene-dependent and environmental malformative influences depends on its whole genetic constitution, which is a mixture of the parental genomes. In placental animals the mother and her offspring together form a unit in which the function and products of two different genotypes come into close contact and have to meet the possible hazards of this disparity. An embryo developing in intimate cellular contact with the mother can be regarded as an allograft, except in inbred strains of animals, because it has a number of antigens inherited from the father which, under ordinary transplantation conditions, would cause immune reactions and rejection if transferred to the maternal organism.

The zygote is in a genetically alien environment right from the moment of fertilization, but the first permanent cell-to-cell contact between embryonic and material tissues is established with the implantation of the blastocyst into the uterine wall. Why then, does an immunologic graft reaction not occur?

Previously it was believed that the cells of the blastocyst do not possess transplantation (also called histocompatibility) antigens, but more recent studies performed with mice have shown that this is a fallacy (Simmons and Russel, 1965; Kirby *et al.*, 1966). There may, however, be a relative deficiency of antigens in the trophoblastic layer of the embryo and the uterine wall may also be an immunologically privileged site during the early stages of pregnancy. It has been shown in experiments performed with mice that the antigenic (genetic) differences between the embryo and the mother are not without consequences at the time of implantation and placenta formation (see James, 1967). Paradoxically, antigenic dissimilarity between mother and embryo favors the growth of the placenta and the embryo; this effect may even be enhanced by previous immunization of the mother with paternal antigens (Billington, 1964; James, 1967). The placenta is formed as a result of trophoblastic invasion of the uterine endometrium and the decidual reaction of the endometrium. It has been postulated that this interaction resembles an immune reaction and, being more extensive in incompatible than in compatible implantation, results in the formation of a larger placenta in the former case.

During later development, the absence of an allograft reaction of the mother against the fetus is generally attributed to the barrier function of the placenta (see Billingham, 1964). In functioning as a barrier, the placenta performs a very delicate task because it must simultaneously allow exchange of gases and nutrients, and relatively large molecules can pass through. In man, the placenta is even responsible, to a large extent, for the immunologic protection of the fetus by passing through immunoglobulins of small molecular weight and antibodies belonging to this group (IgG). On the other hand, the placenta has to separate the fetus and mother in such a way that they will not develop destructive immune reactions to each other and it is to be expected that in this respect it will sometimes fail. The development of the immune mechanism in the embryo and the immunologic interplay between mother and fetus basic to our understanding of the immunopathology of gestation fall beyond the scope of this discussion and the reader is referred to recent reviews of the subject (for example, Galton, 1967; Sterzl and Silverstein, 1967). We will only discuss briefly one example of the immunopathology of gestation, showing the consequences of genetic incompatibility of the mother and her offspring in a situation where the genotype of mother and fetus are normal in the ordinary sense of the word. About 15 percent of Caucasians do not carry the Rhesus (Rh) antigen D and if a Rh-negative mother carries a Rh-positive fetus she may be immunized by fetal erythrocytes carrying the D-antigen that pass through the placenta from fetus to mother. During subsequent pregnancies with a Rh-positive fetus, the mother is liable to develop circulating antibodies against the D-antigen; these are transferred to the fetus and cause a well-known disorder, erythroblastosis fetalis, in the offspring. Other red cell antigens, belonging to the ABO and other blood group systems may also be passed from fetus to mother and there is evidence that some combinations of ABO incompatibility cause embryonic death and early abortion in high frequency.

An interesting situation arises when there is incompatibility of both ABO blood groups and Rh antigens between mother and fetus.

In this case the chances of maternal immunization against Rh antigens is decreased because natural antibodies (isoagglutinins) against ABO antigens in the maternal organism eliminate the fetal blood cells before they can produce an immune response against Rh antigens (Levine, 1943). The understanding of the immunization mechanism underlying erythroblastosis and the beneficial effect of antibodies capable of eliminating fetal red cells has now enabled clinical therapy to be applied on a large scale. Rh-negative mothers pregnant with a Rh-positive fetus are passively immunized with large amounts of human anti-D antibodies. These eliminate any fetal red cells with D-antigen that have escaped to the mother. Thus active immunization of the mother against Rh-antigen is prevented (Clarke *et al.*, 1966).

How often other antigens, cellular and humoral, that have escaped from fetus to mother and vice-versa, cause immunization and the consequences of such immunization is still a matter for speculation (see Brent, 1966; Galton, 1967). On the other hand, it is known that antibodies against fetal and adult tissues from another species injected into a pregnant animal cause abnormal development and congenital anomalies in the embryo (see Brent, 1966).

We may ask now why, in the long course of evolution, which basically means a change in the genetic constitution of a species, the genes responsible for health hazards are not selectively eliminated from the population. It may be postulated that genes responsible for fetal wastage or congenital disease are usually selected out even if the disadvantage compared to the noncarrier population is slight or moderate. Sometimes, however, heterozygosity for a harmful recessive gene represents an advantage for the survival of the population and thus so-called balanced polymorphism is established. Among the few cases known in man, the gene for abnormal hemoglobin S may serve as an example of the advantage of heterozygosity. In a homozygote all hemoglobin is of S-type and leads to a hemolytic disease, sickle-cell anemia, which terminates fatally during adolescence. In heterozygotes some 30–40 percent of the hemoglobin is abnormal, but the hemolytic disease does not develop. The existence of the sickle cell gene is common in areas where falciparum malaria is extremely prevalent, particularly in Africa. In some of these areas up to 40 percent of the population may be carriers of the gene. Heterozygotes for the sickle cell gene are more resistant to malaria than people with entirely normal hemoglobin. Therefore, a balance is established between the existence of genes for sickle-cell anemia and of those for normal hemoglobin by death of children with anemia and malaria, respectively.

It may be considered that a proportion of the congenital defects and perinatal losses due to genetic polymorphism is the price that the population has to pay for possessing a genetically adaptive system against environmental hazards (Neel, 1961).

SUGGESTED READINGS

Review Articles

Auerbach, C. The chemical production of mutations. Science **158**: 1141–1147 (1967).

Bartalos, M. and T. A. Baramki. Medical cytogenetics. Williams & Wilkins Co., Baltimore (1967).

Billingham, R. E. Transplantation immunity and the maternal-fetal relation. New Engl. J. Med. **270**: 667–672, 720–725 (1964).

Brent, R. L. Immunologic aspects of developmental biology. *In* Advan. Teratol., (D. H. M. Wollam ed.), vol. I, pp. 81–129. Logos Press, London, Academic Press, New York (1966).

Carter, C. O. The inheritance of common congenital malformations. *In* Progr. Med. Genet., Vol. IV. ed. by (A. G. Steinberg and A. G. Bearn), vol. IV, 2nd ed., pp. 59–84. Grune & Stratton, New York and London (1965).

Day, R. W. The epidemiology of chromosome aberrations. Symp.: Etiology of chromosomal abnormalities. Am. J. Human Genet. **18**: 70–80 (1966).

Federman, D. D. Abnormal sexual development. W. B. Saunders Co., Philadelphia and London (1967).

Galton, M. Immunological interations between mother and fetus. *In* Comparative aspects of reproductive failure, (K. Benirschke, ed.) pp. 413–446. Springer-Verlag, Berlin, Heidelberg, New York (1967).

Grüneberg, H. The pathology of development. A study of inherited skeletal disorders in animals. Blackwell, Oxford (1963).

Inhorn, S. L. Chromosomal studies of spontaneous human abortions. *In* Advan. Teratol., (D. H. M. Woollam, ed.) vol. 2, pp. 37–99. Logos Press, London, Academic Press, New York (1967).

Levine, R. P. Genetics, 2d ed. Holt, Rinehart and Winston, New York, (1968).

McKusick, V. A. Human genetics. (Prentice-Hall Foundations of Modern Genetics Series.) Prentice-Hall Inc., Englewood Cliffs, New Jersey (1964).

Nichols, W. W. The role of viruses in the etiology of chromosomal abnormalities. Symp.: Etiology of chromosomal abnormalities. Am. J. Human Genet. **18**: 81–92, (1966).

Polani, P. E. Chromosome aberrations and birth defects. *In* Birth defects, (M. Fishbein, ed.) pp. 136–155. J. B. Lippincott Co., Philadelphia (1963).

Stanbury, J. B., J. B. Wyngaarden, and D. S. Fredrickson, (Eds.): Metabolic basis of inherited diseases, 2d ed. McGraw-Hill Book Co., New York, Toronto, Sydney and London (1966).

Sterzl, J. and A. M. Silverstein. Developmental aspects of immunity. *In* Advan. Immunology, (F. J. Dixon and J. H. Humphrey, eds.), vol. 6, pp. 337–459. Academic Press, New York and London (1967).

Taylor, A. I. Patau's, Edwards' and Cri du Chat syndromes: A tabulated summary of current findings. Develop. Med. Child Neurol. **9**: 78–86, (1967).

Wagner, R. P. and H. K. Mitchell. Genetics and metabolism. 2d ed. John Wiley & Sons, New York (1964).

Warkany, J., E. Passarge, and L. B. Smith. Congenital malformations in autosomal trisomy syndromes. Am. J. Dis. Childr. **112**: 502–517, 1966.

Watson, J. D. Molecular biology of the gene. W. A. Benjamin Inc., New York, Amsterdam (1965).

Special Articles

Aula, P. Virus-associated chromosome breakage. A cytogenetic study of chickenpox, measles and mumps patients and of cell cultures infected with measles virus. Ann. Acad. Sci. Fennicae. Ser. A IV. Biol. No. 89, 1–75 (1965).

Billington, W. D. Influence of immunological dissimilarity of mother and foetus on size of placenta in mice. Nature **202**: 319–320 (1964).

Buckton, K. E., P. A. Jacobs, W. M. Court Brown, and R. Dall. Study of chromosomal damage persisting after x-ray therapy for ankylosing spondylitis. Lancet **2**: 676–682 (1962).

Carr, D. H. Chromosome anomalies as a cause of spontaneous abortion. Am. J. Obstet. Gynecol. **97**: 283–293 (1967).

Clarke, C. A., C. C. Bowley, J. Shaw, and W. B. Bias. Prevention of Rh-haemolytic disease: results of the clinical trial. A combined study from centers in England and Baltimore. Brit. Med. J. **2**: 907–914 (1966).

German, J. Mongolism, delayed fertilization and human sexual behaviour. Nature **217**: 516–518 (1968).

Hall, B., K. Fredgar, and N. Svenningsen. A case of G monosomy? Hereditas **57**: 356–364 (1967).

Jackson, J. F. and W. P. Ashford. Familial mongolism due to 21/22 chromosome translocation. J. Am. Med. Ass. **200**: 722–724 (1967).

Jacob, F. and J. Monod. Genetic regulatory mechanisms in the synthesis of proteins. J. Mol. Biol. **3**: 318–356 (1961).

James, D. A. Some effects of immunological factors on gestation in mice. J. Reprod. Fertility. **14**: 265–275 (1967).

Kirby, D. R. S., W. D. Billington, and D. A. James. Transplantation of eggs to the kidney and uterus of immunised mice. Transplantation **4**: 713–718 (1966).

Levine, P. Serological factors as possible causes in spontaneous abortions. J. Hered. **34**: 71–80 (1943).

Milne, M. D. Lessons from inborn errors of metabolism. Proc. Roy. Soc. Med. **59**: 1157–1162 (1966).

Neel, J. V. Some genetic aspects of congenital defects. *In* Congenital malformations, (M. Fishbein, ed.), pp. 63–69. J. B. Lippincott Co., Philadelphia and Montreal (1961).

Nusbacher, J., K. Hirschhorn, and L. Z. Cooper. Chromosomal abnormalities in congenital rubella, New Engl. J. Med. **276**: 1409–1413 (1967).

Landauer, W. Observations on penetrance and expressivity of Lamoreux's chondrodystrophy of fowl. J. Heredity **56**: 209–214 (1965a).

———. Gene and phenocopy: Selection experiments and tests with 6-aminonicotinamide. J. Exp. Zool., **160**: 345–354 (1965b).
Simmons, R. L. and P. S. Russell. Histocompatibility antigens in transplanted mouse eggs. Nature **208**: 698–699 (1965).
Waddington, C. H. Tendency towards regularity of development and their genetical control. *In* International Workshop Teratol. pp. 66–75, Copenhagen (1966).

4

Genesis of Congenital Defects

By definition, maldevelopment is a divergence from the normal course of development and this may be expected to take place at any step in the sequence of processes leading to the formation of a fully developed organism. The primary code in the genome might be "misleading" or defective, the transcription of this information to the synthesizing machinery could be inhibited, or finally ribosomal protein synthesis might be affected. At the cellular and tissue levels, movements of cells and organ anlagen can be prevented, their interactive processes can be affected, or their growth may be inhibited. Excessive loss of tissue may cause permanent defects or, vice versa, failure of cells to degenerate may lead to abnormalities—these are only a few of the possible mechanisms involved in maldevelopment.

There are examples of malformations for which the genesis can be traced back to a single divergence at a certain stage of development, but more often the end result, whether morphologic or functional, is a consequence of sequential maldevelopment, where failure in one step is followed by a series of subsequent distortions. And conversely, a certain type of malformation may be the result of a variety of failures all leading to the same end result, although via different paths.

MORPHOGENETIC MOVEMENTS

In addition to the random motility of cells in a developing organism, there are many well known and thoroughly analyzed examples of guided, oriented migrations of cells and cell clusters leading to a constant rearrangement of tissue components inside an embryo.

The guiding principles recently have been elucidated in several model systems for such morphogenetic movements, and we are beginning to understand some of the cell surface characteristics apparently responsible for such synchronized migrations (see DeHaan, 1964; Trinkaus, 1965). Because of their random motility, cells constantly make contacts with like cells or with cells that have had a different developmental history and they are able to "recognize" the former, to which they adhere and so establish permanent contact. Such cells, furthermore, are able to sort themselves out and form homogenous colonies in a mixed cell population *in vitro*. Similar selectively adhesive forces seem to guide the oriented movements of larger cell clusters or colonies—here probably the best analyzed example is to be found in the gastrulation of sea urchin embryos (Gustafson and Wolpert, 1963). In this study the mesodermal cells of early embryos send long filamentous processes across the coelomic cavity and "feel" the inner surface of the ectoderm. Finally, these become attached to certain points in the wall and conduct the mesodermal cells toward these points, with the obvious consequence of gastrulation.

These directed movements naturally play a significant role in embryogenesis and severe abnormalities might well be anticipated to result from changes in the surface characteristics regulating them. Such distortion would also be expected to have profound effects on the neighboring tissues known to be dependent on normal interactive processes (see below) and the end result might thus comprise a whole series of failures in normal synchronized development. Defects in the mechanisms directing morphogenetic movements will be exemplified by the following three well-analyzed events.

Gastrulation

During the well-known process of Amphibian gastrulation, the cells of the blastopore lip start an oriented migration toward the interior of the gastrula and glide smoothly in between the ectoderm (presumptive neural plate) and the endoderm to form the mesodermal mantle. Various treatments inhibit this migration and lead to what is known as "exogastrulation." Here a part of the mesoderm or the whole blastoporal lip fails to invaginate and consequently no mesoderm will underlie the presumptive neural plate. The fatal consequences of this inhibition are seen in Fig. 4.1 and they will be analyzed in the section headed "Tissue Interactions," page 85.

Movements of Precardiac Mesoderm

The vertebrate heart is formed from a pair of anlagen which fuse in the ventral midline to become a tubular heart primordium.

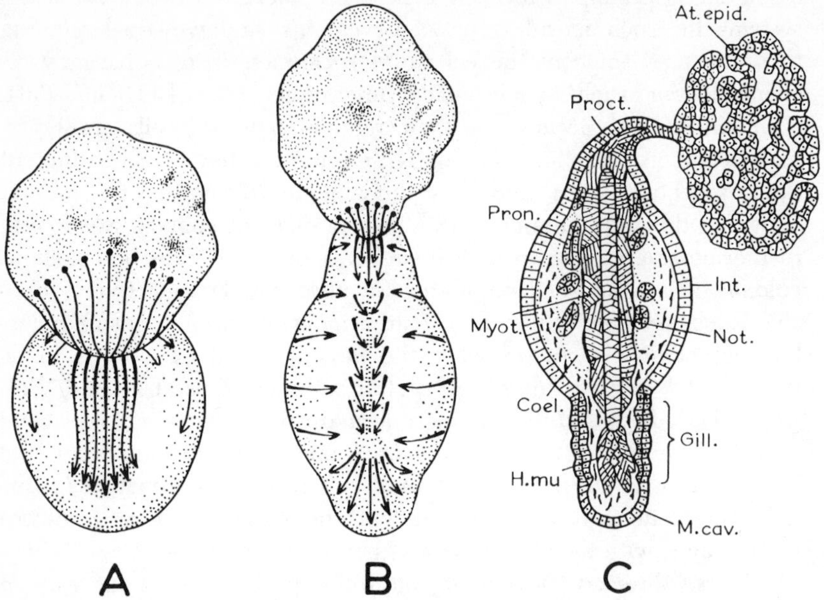

Fig. 4.1 Exogastrulation in amphibian embryo caused by treatment with hypertonic culture solution. A and B illustrate two successive stages of the process and C shows the anatomy of the resulting monster *At. epid.*, atypical epidermis; *Coel.*, coelomie cavity; *Gill.*, gill epithelium; *M. cav.*, mouth cavity; *Myot.*, myotomes; *Not.*, notochord; *Proct.*, protodeum; *Pron.*, pronephric tubules. (From L. Saxén and S. Toivonen. Primary Embryonic Induction, Academic Press, London [1962].)

In the chick the origin of these anlagen has been traced back to early stages of development and they have been shown to originate at the head process stage from the area around the anterior end of the primitive streak. Their migration from here to the midline represents an oriented, conducted process, which has been thoroughly analyzed by time-lapse cinematography by DeHaan (1963, 1964). In the beginning, their movement is random but it soon becomes oriented and the cell clusters begin to glide toward the site of heart formation. The exact nature of the forces guiding this migration still is not fully understood, but experimental evidence strongly suggests that the endodermal substrate of the migrating clusters provides a path along which the cells move. The mesodermal cells are known to send out filamentous probes—as in the sea urchin gastrula—and these may perform the same explorative function in finding the path leading to the heart-forming area. Prevention of this directed movement would be expected to inhibit the fusion of the cardiac anlagen and so lead to the development of a paired heart, known from earlier descriptions as "cardia bifida." On the basis of the known role of divalent cations in establishing cellular contacts, DeHaan made experiments in which explanted chick embryos were treated with chelating agents that removed these cations from the environment. The results illustrated in Figs. 4.2 and 4.3 show that the chelating agent EDTA (ethylenediamine tetraacetic acid) in a concentration around 5 mm does, in fact, lead to a "cardia bifida" malformation. In the same concentration these first signs of a generalized disaggregation were noted, as indicated in the curve, whereas concentrations ineffective in producing cardiac maldevelopment did not show such effects on the embryo. This correlation strongly suggests that we are dealing with disturbances in cellular contacts and mutual adhesiveness. The role of Ca^{++} ions was subsequently confirmed by adding equimolar concentrations of EDTA and calcium with no apparent consequences of EDTA concentrations known to inhibit cell migration in the absence of Ca^{++}. All this suggests that a disruption of certain cellular contacts had prevented the directed movements of the precardiac clusters. In view of the apparent guiding role of the endodermal substrate, it is conceivable that this disruption of contact has taken place between the migrating mesodermal cells and the underlying endoderm.

Fig. 4.2 Dose response curve of EDTA in a calcium-free culture medium. The histogram indicates the percentage of cardia bifida, the curves the proportion of viable embryos (○) versus embryos in poor condition showing signs of generalized disaggregation, (●). (After R. L. DeHaan. *In* The Chemical Basis of Development, McElroy and Glass [eds.], p. 339. The Johns Hopkins Press, Baltimore [1958].)

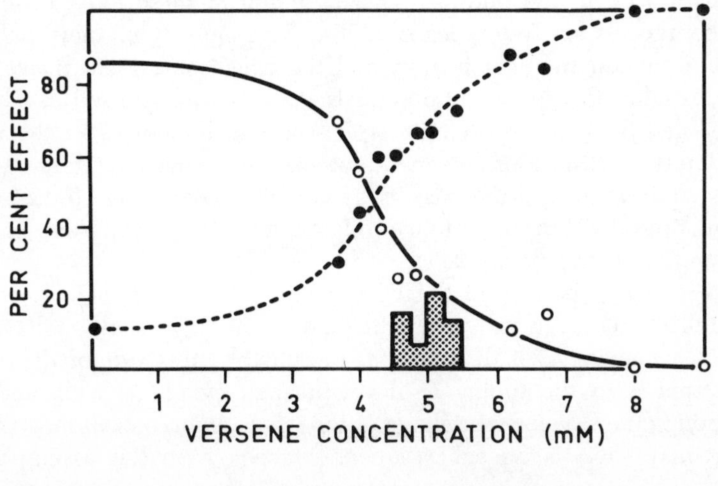

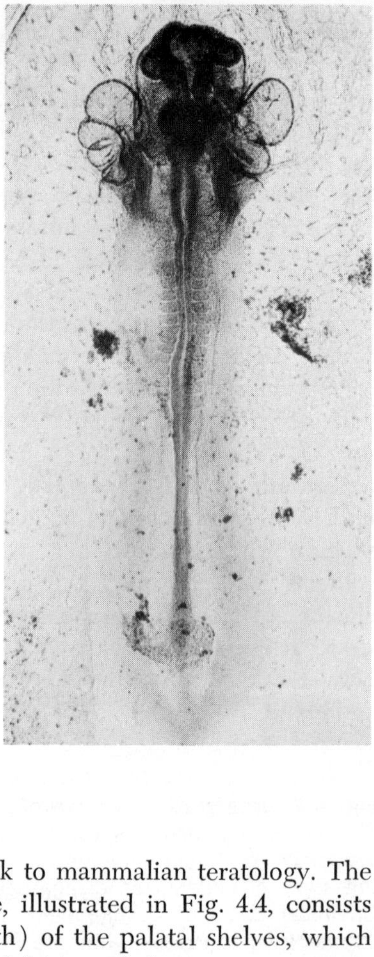

Fig. 4.3 Cardia bifida anomaly in explanted chick embryo produced by treatment with a chelating agent. (Courtesy of Dr. Robert DeHaan.)

Closure of the Palatal Shelves

Our third example brings us back to mammalian teratology. The development of the secondary palate, illustrated in Fig. 4.4, consists of a series of movements (and growth) of the palatal shelves, which normally fuse in the midline. The inhibition of these movements or of the process of fusion leads to the development of cleft palate, a malformation frequent in man and the object of a great many detailed studies in experimental animals. As has been stressed in Chapters 2 and 3, there are definite strain-specific differences in the susceptibility of this morphogenetic process to exogenous teratogens. These differences are the basis of remarkably interesting studies and calculations by Fraser and his collaborators, to be reported below (summarized by Fraser, 1965).

Fraser first points out that, irrespective of what biologic measure we follow, there are always individual variations, usually following the Gaussian curve of distribution. Presumably, this sort of distribution applies to the timing of developmental events as well, and in relation to the developmental age of the whole embryo each individual event may have its typical variations. Starting from this assumption,

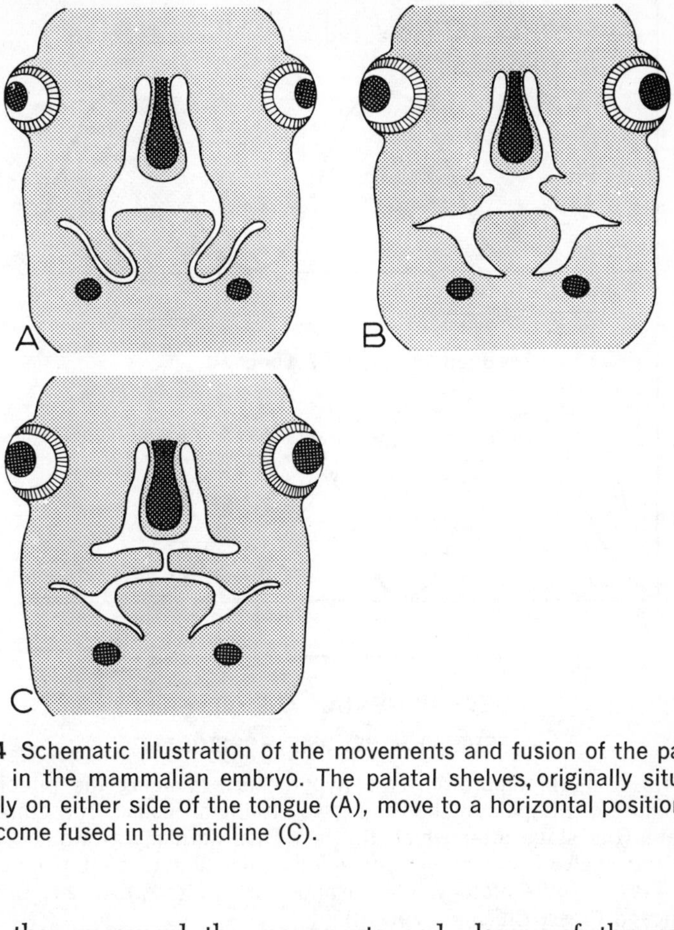

Fig. 4.4 Schematic illustration of the movements and fusion of the palatal shelves in the mammalian embryo. The palatal shelves, originally situated vertically on either side of the tongue (A), move to a horizontal position (B) and become fused in the midline (C).

Fraser then measured the movements and closure of the palate. When comparison was made between two inbred mouse strains, A/J and C57BL, it became evident that closure of the palate took place later in the A-strain than in the C57BL mice. Knowing from earlier experimental work that strain A/J is highly susceptible to cortisone, which produced cleft palate in 100 percent of the mice (Table 2.1, page 20), in contrast to the C57BL strain, in which palatal defects could be induced in only 20 percent of the embryos, Fraser concludes with the hypothesis illustrated in Fig. 4.5.

This means that the palatal shelves are able to fuse and form a complete secondary palate only up to a certain stage of development of the embryo. Beyond this threshold, growth of the head has advanced to a stage where the cleft to be covered by the shelves is too wide to allow contact between them. Treatment with a teratogen now may cause retardation of the movements of the shelves and

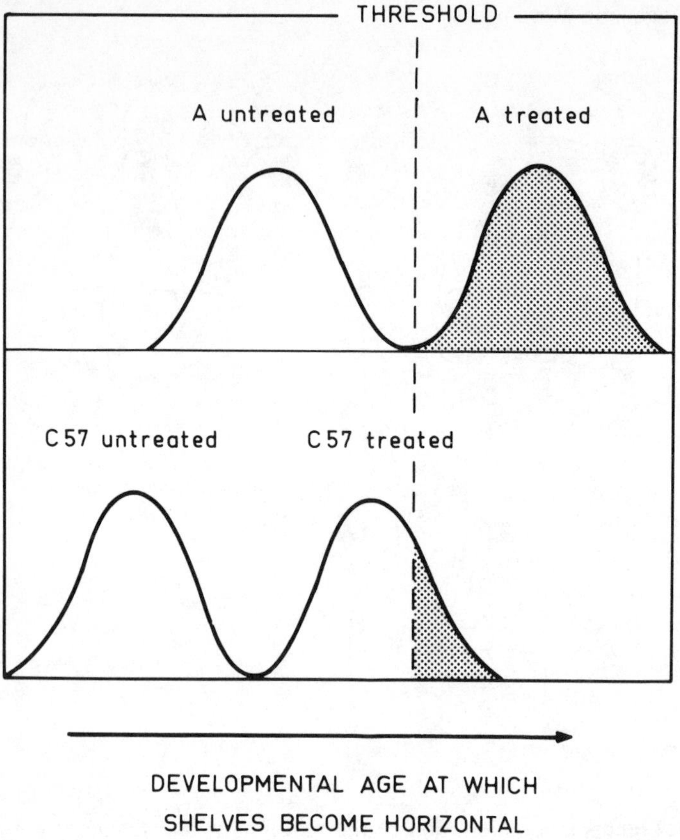

Fig. 4.5 Diagram illustrating the hypothesis explaining the inhibition of the closure of the palatal shelves in two strains of mouse. The "threshold" indicates the stage after which the head has normally grown too wide to allow the shelves to meet in the midline, leading to cleft palate. (After F. C. Fraser. *In* Teratology, Wilson and Warkany (eds.), p. 21, University of Chicago Press, Chicago [1965].)

bring a certain number of the embryos beyond this threshold. In other words, the whole distribution curve is shifted toward a more advanced stage of development of the embryo, the oldest end of the curve being pushed over the threshold stage. With the knowledge that closure occurs earlier in normal C57BL mice than in A/J strain embryos, the differences in their responsiveness to exogenous teratogens (cortisone) can be well explained in terms of this hypothesis. Moreover, the additive effect of several simultaneous or even successive teratogens (Kalter, 1965; Runner, 1965) may be explained in Fig. 4.5. Each simultaneous teratogen may exert its own retarding effect by shifting the closure process toward older stages and therefore bringing more and more embryos beyond the threshold, resulting in formation of a defective palate.

Evidence supporting this hypothesis has recently been gained from *in vitro* studies, where mouse palates were cultivated *in vitro* as described on page 29. In the presence of hydrocortisone, their growth and fusion were clearly delayed (showing the retarding effect of the drug), but complete fusion was never inhibited (indicating lack of the threshold probably determined by the *in vivo* growth of the head) (Lahti and Saxén, 1967).

TISSUE INTERACTIONS

On account of the morphogenetic movements, a continuous rearrangement of cells and tissues takes place inside a developing embryo and consequently, the micromilieu of differentiating cells is constantly changing. There are a great many examples demonstrating the morphogenetic significance of this changing enviroment and the role of tissue interactions in normal, synchronized development. To define such interactive or "inductive" processes, we may quote Grobstein (1956): "inductive tissue interaction takes place whenever in development two or more tissues of different history and properties become intimately associated and alteration of the developmental course of the interactants results."

Realization of the importance of such interactions in normal development prompts a search for failures in these inductive processes as mechanisms underlying maldevelopment. There are, in fact, many examples which convincingly demonstrate that in several genetically determined syndromes the developmental abnormality depends on failure of the inductive system. Examples in which an exogenous teratogen has led to inhibition or distortion of inductive interactions are less numerous, but experimental embryology can indicate in different ways how this mechanism may be affected. However, before any of these well-analyzed examples are presented, we should briefly recall certain basic features of inductive interactions. This has been done below in the light of "primary induction," the classic model of the developing central nervous system, which is one of the best analyzed inductive systems. It should be stressed, however, that the whole sequence of events termed "induction" is still not sufficiently well known to warrant generalizations or to allow formulation of a general scheme of the process applicable to different interactive systems.

Primary Induction

The series of morphogenetic movements leading to the inductive interactions responsible for the normal segregation of the central nervous system is the process of gastrulation. If gastrulation is prevented experimentally, the inductive blastopore lip does not come into contact with the responding presumptive neural plate and consequently no central nervous system develops. Figure 4.1 indicates the situation after total exogastrulation, where the ectoderm remains as atypical, undifferentiated tissue. This simple experiment demonstrates convincingly the central role played by induction (and naturally by the preceding morphogenetic movements). Some of the present conceptions of the inductive processes during and after gastrulation will be summarized in Fig. 4.6.

During early gastrulation, an inductive stimulus is released from the invaginating blastopore lip (I) and this triggers the differentiation of the overlying ectoderm. At this stage the ectoderm is still undetermined—"totipotent," that is to say, with practically unlimited developmental capacities (experimentally, the ectoderm at this stage can be induced to differentiate not only to neural and neuroectodermal derivatives, but to mesodermal and endodermal components as well). This responsiveness to certain inductive stimuli is generally referred to as competence (C_I). Triggering of differentiation leads to a partial restriction of this competence and the differentiative capacities of the corresponding tissue become limited; even before any overt morphogenesis can be observed, the competence to respond to the primary inductors is lost. Subsequently, the predetermined tissue is exposed to secondary interactive influences (II) and owing to the lability

Fig. 4.6 A scheme of the processes interacting during the determination and subsequent segregation of the central nervous system.

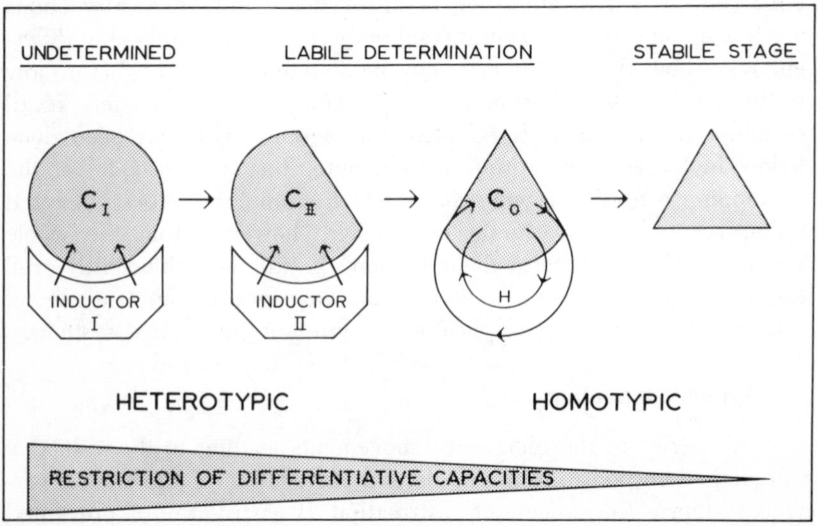

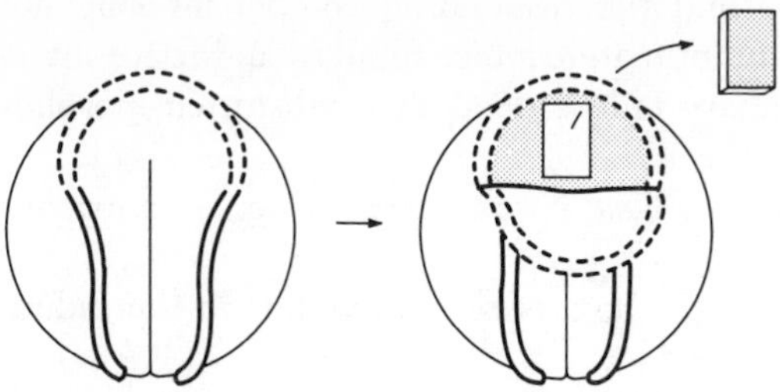

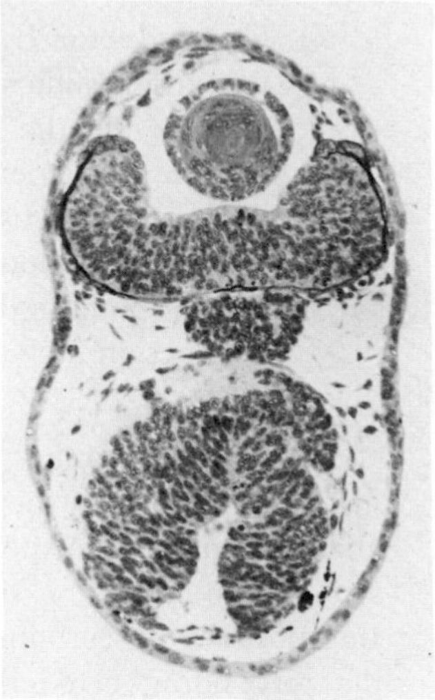

Fig. 4.7 Experimental production of cyclopy in the amphibian embryo. Removal of a central zone of the head region of the inductive prechordal plate results in narrowing of the inductor tissue and, in consequence, fusion of the eye vesicles occurs. (*See* O. Mangold. Anat. Anz., Suppl. p. 3 [1958].)

of the determination its differentiative path can be altered by such secondary inductions, although not within the wide range mentioned above for earlier stages (the competence is already restricted, C_{II}). Finally, the competence to react to such "heterotypic" stimuli is lost (C_0) but the ectodermal cells are still dependent on their environment, now consisting of like cells. Such "homotypic" interactions between like cells are known to be of the greatest significance in the maintenance of the differentiative state resulting from a response to inductors of different orders.

For a direct demonstration of the role of this inductive system in teratogenesis, an operative manipulation for the production of cyclopy is illustrated in Fig. 4.7. When a portion of the anterior part of the inductor tissue is removed at the early gastrula stage, the template triggering the determination of the overlying ectoderm becomes spatially restricted. As a consequence, the induced area in the intact ectoderm is narrowed and the lateral optic vesicles are brought close together. According to the extent of the primary operation, this leads to partial fusion of the eyes or to complete cyclopy, as in the case illustrated in Fig. 4.7. This same experiment may serve as an example of a distorted secondary induction: the lens is formed from the overlying epidermis after an inductive trigger is released by the optic vesicle. In a partially fused eye, we usually see two small lens vesicles in close proximity and in the case illustrated above, the abnormal cyclopic eye has induced only one large lens.

On the basis of our knowledge of primary induction and of the information gained from experimental interference with this developmental mechanism, three types of failures in inductive tissue interactions can be postulated:

1. The inductor tissue and the responding component may not come into sufficiently intimate contact to allow inductive interaction (and the primary failure depends on distorted morphogenetic movements).
2. The inductor tissue has lost its capacity to act as a trigger of differentiation.
3. The responding tissue has lost its responsiveness to the inductive action.

Before presenting known examples of these alternative failure mechanisms, it may be stated that all these can easily be produced and demonstrated experimentally in the model system already presented, the induction of the central nervous system in amphibians. Interposition of certain materials between the archenteron roof and the responding ectoderm prevents induction, different physical and chemical treatments of the inductor tissue abolish its inductive properties, and aging of the presumptive neural plate region outside the embryo leads to rapid loss of its responsiveness to inductive stimuli. This last mentioned and repeatedly made observation of the temporal restriction of competence might, in fact, lead us to take the time factor as a fourth possibility and indicates that as a consequence of delayed morphogenetic movements, the inductor tissue may not come into contact with the responding elements until the elements have already lost their competence. In other words, a synchrony of the interactive processes is necessary for normal development. (For further details on primary induction see Holtfreter and Hamburger, 1955; Saxén and Toivonen, 1962.)

Failure To Make Contact

Three examples of normal inductive interactions have so far been mentioned here: induction of the central nervous system, interaction leading to the formation of the lens, and an epitheliomesenchymal interaction involved in kidney development (page 31). Examples showing defects caused by failure of intimate contact between the inductor and the responding tissue can be shown in all three systems, the first of which was already represented by exogastrulation. Lack of proper contact between the optic cup and the overlying ectoderm seems to be the failure responsible for anophthalmy in mice as described by Chase and Chase (1941). They have given a detailed analysis of the developmental events preceding the eyelessness (or microphthalmia) occurring in a certain strain of mice. Optic vesicles are formed, but do not make contact with the overlying epidermis, and as a result no lens is induced. In cases where some degree of contact is achieved, a small lens is formed, demonstrating that the epidermis is competent to respond to the proper inductive stimulus of the eye and that the latter is retained in the strain. The anomaly can thus be traced back to the lack of intimate contact between the interactants. (It may be stated here that by "intimate contact" we do not necessarily mean a direct cell-to-cell contact. It has been shown in the lens-optic vesicle system, as well as in other tissue interactions, that interposition of porous material between the interactants does not prevent induction.)

Our third example refers to the induction of kidney tubules, which was already discussed on page 31. There we learned that the development of the secretory tubules in the metanephrogenic mesenchyme is triggered by an inductive stimulus from the ureteric bud and that this interactive system can be submitted to *in vitro* analysis. In the mutant *Sd* mice, the homozygote *Sd/Sd* completely lacks kidneys and excretory openings whereas the heterozygote *Sd/+* shows less severe malformations of the urinary tract. An *in vitro* analysis performed by Gluecksohn-Waelsh and Rota (1963) indicated, however, that the mutant *Sd* did not completely suppress the inductive capacity of the ureteric bud or the competence of the metanephrogenic mesenchyme to respond to the inductive stimulus. Reciprocal combinations with the interactants from a normal mouse resulted in almost normal *in vitro* development, although quantitative differences could be noted. It is not possible to conclude with certainty where in this case the interactive process has been affected by the mutation and there are at least two alternative explanations: either the inductive capacity and/or the competence have been *partially* suppressed by the mutation; or the development and branching of the ureteric bud is retarded, so that the normal course of synchronized development is confused. Here, as perhaps in the case of otocephaly, the time factor may play an important role, as the inductor may not reach the responding tissue until after this period of competence has ended.

Loss of Inductive Capacity

Before we can cite examples of congenital defects where the failure can be traced back to an abnormal inductive capacity of tissues, a new interactive system has to be described. In the three induction systems already referred to (central nervous system, lens, kidney), the interaction was somewhat schematically considered to be one-sided, where one of the interactants served as inductor and the other represented a purely responding component. In fact, this is an oversimplification and in all the interactive systems studied so far, the process is more or less mutual in character. In the interactive processes leading to the development of the vertebrate limb, this reciprocal nature is quite clear and rather well analyzed, as will be seen in the following summary based mainly on the studies of Saunders (1948) and Zwilling (1956).

The limb bud consists of two components: a mesodermal condensate and an overlying ectoderm. At the beginning of the whole series of interactive processes the mesodermal condensate induces the formation of a thickened ectodermal ridge, which in turn acts as an inductor on the mesodermal blastema. The latter induction is, in fact, a continuous series of interactive processes, where the proximal parts are first induced by the ectoderm and subsequently new, more distal zones are determined and added to the oldest proximal portion. This sequence can be demonstrated by removing the ridge at a stage where only the proximal parts have been determined—the operation results in an abnormal limb in which only the femur is present. If, on the other hand, the whole distal part of the mesodermal blastema is removed, but the remaining proximal part is covered by ectoderm from the same stage as in the foregoing experiment, a complete limb will be induced (Fig. 4.10).

Furthermore, the ectoderm and its inductive capacity are not independent of the mesoderm, which provides a maintenance factor responsible for the function of the ectoderm. This factor is asymmetrically distributed in the mesoderm and labels a corresponding area in the ectoderm typical of each type of limb. This can be demonstrated by the experiment illustrated in Fig. 4.8: when reciprocal combinations of ectoderm and mesoderm from normal and polydactylous limbs were made, the resulting shape and type was determined by the mesoderm, despite the fact that the actual inductor is known to be localized in the ectoderm (as shown above).

Let us now return to the title of this section, which suggests examples of abnormal inductive capacity of tissues. The studies of Zwilling (1956) on the wingless mutation in the chick provide a well-analyzed example of a situation where an inductive capacity has been lost. The homozygous recessives of this mutant strain fail to form wings, but embryologic analysis has shown that the limb bud develops normally until the beginning of the third day. Thereafter the apical ectodermal ridge degenerates and the development of the whole bud ceases. An experimental combination of mutant and normal limb bud components gives interesting results (Fig. 4.9). As might be expected, the combination of mutant mesoderm with mutant ectoderm resulted in failure to develop and the same observation was

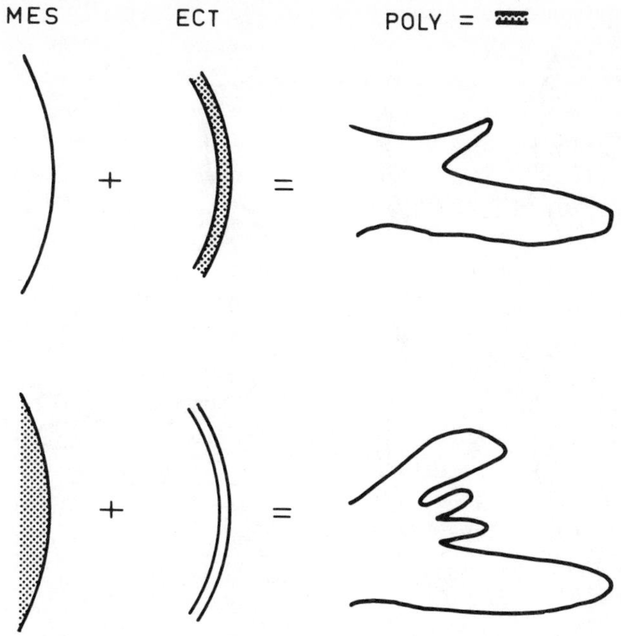

Fig. 4.8 Interchange between normal and polydactylous limb buds, demonstrating the decisive role of the mutant mesodern. (After E. Zwilling. Cold Spring Harbor Symp. Quant. Biol. **21:** 349 [1956].

made when normal mesoderm was brought into contact with mutant ectoderm. However, when mutant mesoderm and normal ectoderm were brought into interactive contact, the limb bud showed more distal differentiation apparently induced by the normal ectoderm. The differentiation was not completed, however; the apparent reason was lack of a maintenance factor in the mutant mesoderm leading to regression of the inductive ectodermal ridge. The results thus suggest that the mutant mesodermal blastema is competent to respond to the morphogenetic stimulus of the ectoderm and that the defect leading to inactivation of the inductor is in its maintaining properties.

The interactive processes in the limb bud and the information gained from the above investigations are of the greatest interest when the effects of certain exogenous teratogens on limb development are tested. This has recently been done, especially by Etienne Wolff and his colleagues (see Wolff, 1966). For the understanding of their results, the proximodistal sequence of the induction by the ectodermal ridge should be borne in mind and, hence, the experimental evidence (Fig. 4.10) will be repeated. If the ectodermal ridge is removed

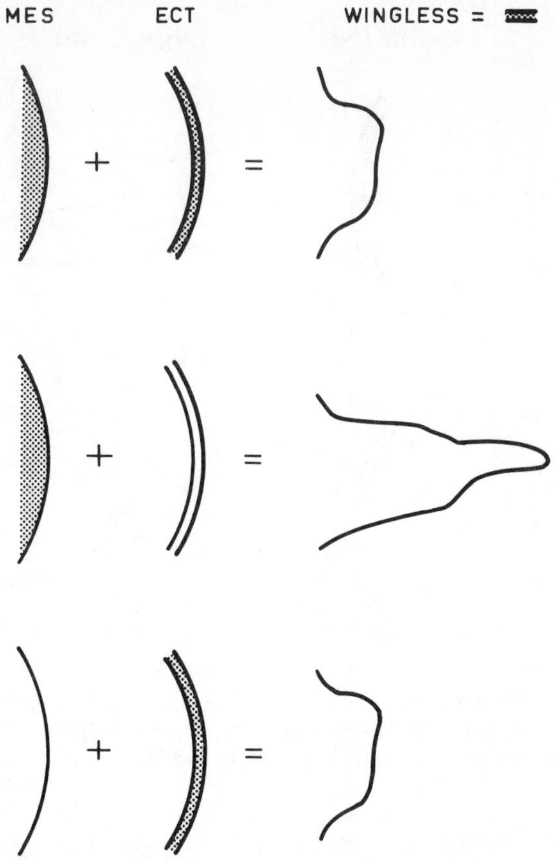

Fig. 4.9 Interchange between normal and wingless limb buds demonstrating the "competence" of the mutant mesoderm and its defective maintenance properties. (After Zwilling Cold Spring Harbor Symp. Quant. Biol. **21**: 349 [1956].)

from a young limb bud and subsequently replaced by a new ectoderm cap from an embryo in the same stage, a complete limb will be induced (as would be expected). If, on the other hand, the decoated blastema is brought into contact with an ectodermal cap from an older embryo, only the distal parts of the limb will be induced, because at this stage the ectodermal ridge has already lost its capacity to determine the development of the proximal parts.

Treatment of the limb bud with nitrogen mustard, a potent teratogen, leads to severe malformations of the phocomely type (Fig. 4.11) and the explanation of the mechanism by which this anomaly is determined may be found in the recent observations of Wolff and his collaborators. Histologic examination of the treated limb buds has

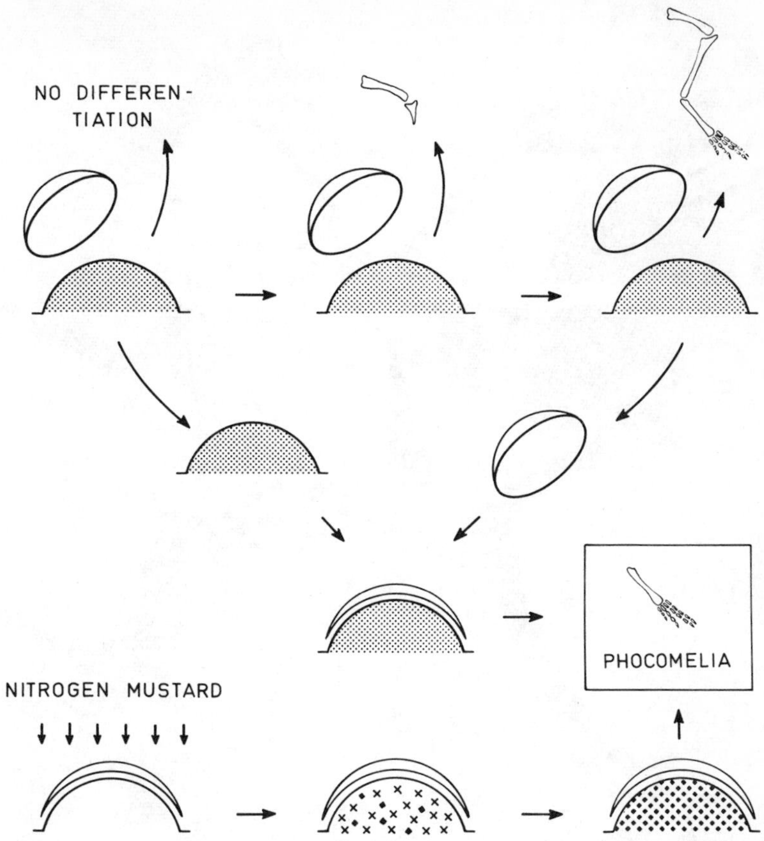

Fig. 4.10 Transplantation experiments on the chick limb bud and the effect of nitrogen mustard on the limb developed, schematically presented. (Based on experiments and conclusions by Et. Wolff, B. Salzgeber, A. Hampe and M. Kieny; *see* Wolff [1966].)

indicated that nitrogen mustard exerts its primary action on the mesodermal component of the bud, in which profound degenerative changes are seen some 20 to 40 hours after treatment. Whereas a great part of the mesodermal blastema is lost during this period, the ectodermal ridge is apparently undamaged and shows no signs of cell death. After 72 hours the mesoderm shows signs of recovery and the cells which survived the treatment begin to proliferate and differentiate under the inductive control of the viable ectodermal ridge. But the latter has meanwhile aged and lost its capacity to induce proximal structures of the limb and consequently the recovering mesoderm is exposed to a restricted inductive action leading to the determination of distal parts only. As a result, a limb develops in which only the digits and, perhaps metacarpals (metatarsals) can

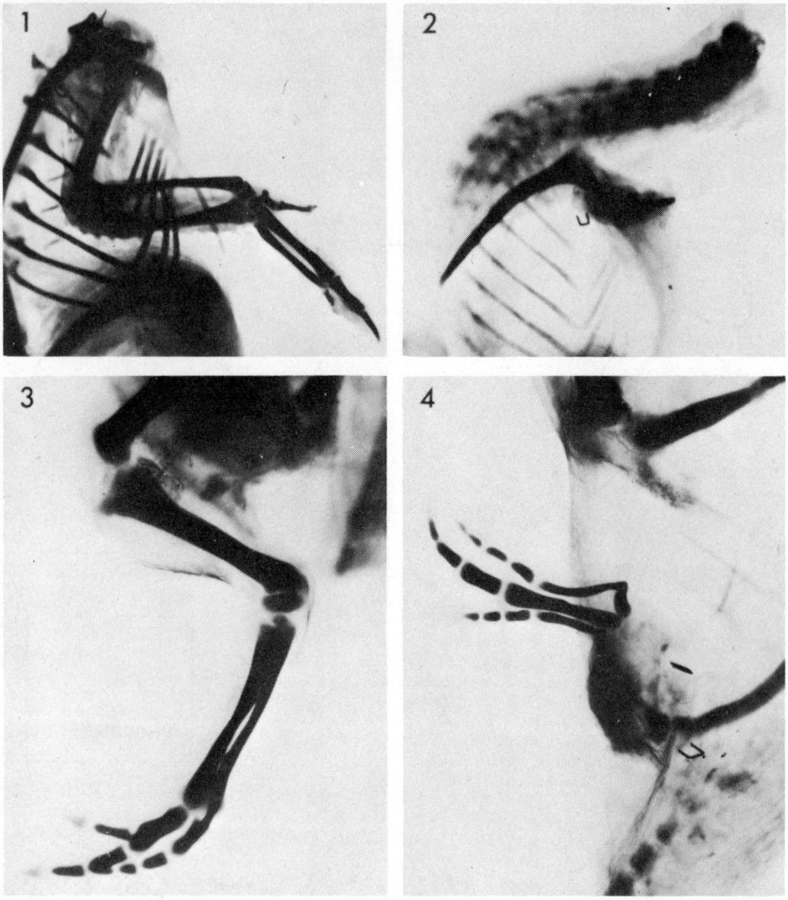

Fig. 4.11 Results obtained by treating limb buds of chick embryos with nitrogen mustard. 1 and 3, wing and leg of control embryos; 2 and 4, phocomely as a result of nitrogen mustard treatment respectively. Cleared and alizarin strained preparations. (Parts 1 and 2 from B. Salzgeber. C. R. Acad. Sci. Paris, **264:** 395 [1967]. Parts 3 and 4 courtesy Dr. B. Salzgeber.)

be discerned. This series of experiments strongly suggests a mechanism in which both a direct effect on responding cells in an interactive system and a subsequent confusion of normal sequential induction are involved and lead to maldevelopment. The experiments described above are summarized in Fig. 4.10 and some of the results illustrated in Fig. 4.11.

Loss of Competence

The term "competence" has been defined by Holtfreter and Hamburger (1955) as the "physiological state of a tissue, which permits it to react in a morphogenetically specific way to determinative stimuli." Consequently, loss or restriction of this responsiveness would undoubtedly lead to developmental defects, as can easily be demonstrated experimentally. When the presumptive neural plate is removed from a young gastrula (before induction), cultivated *in vitro* for 24 hours, and subsequently exposed to an inductive stimulus, no differentiation follows. The experience gained from such experiments shows that the competence in the aging ectoderm is gradually and selectively lost, leading first to a restriction of the repertoire of differentiations and soon to a complete unresponsiveness to inductive stimuli of the first order (Fig. 4.6). Yet the cells remain viable and may respond to certain inductive stimuli of the second order, as indicated in the scheme. Experiments, still mainly on amphibian embryos, have furthermore shown how such changes in competence can be effected with a variety of exogenous factors such as chemicals, irradiation, pH, and so forth (see Saxén and Toivonen, 1962). Thus, at least three different mechanisms may be postulated by which competence could be altered and so play a significant role in maldevelopment:

1. Competence may be genetically restricted.
2. Exogenous teratogens may primarily affect competence, while leaving the inductor unaltered.
3. Retarded morphogenetic movements may prevent the interactants from coming into contact until after the temporarily restricted period of competence.

The two latter alternatives have repeatedly been demonstrated by experimental embryologists (above), and future teratologic analyses will certainly provide further examples of them. As an example of genetic restriction of competence, a mutation of the T-locus in mice should be discussed. Before presenting the interesting observations of Dunn and Bennet (summarized 1964), some comments on the induction of cartilage in somite tissue should be made (see Holtzer, 1964; and Lash, 1963). The formation of cartilage by the somite tissue is still another example of tissue interaction and a potent chondrogenic inductor seems to be the ventral spinal cord of young embryos. The experimental setup for the demonstration of this induction is illustrated schematically in Fig. 4.12 and has been employed in the analysis of the defects in the T-locus mutants. In these mice, the principal defect is failure of development of the vertebral cartilages, which are either not developed at all or abnormally developed. An *in vitro* analysis following the experimental design of Fig. 4.12 and 4.13 showed that the spinal cord tissue of the mutant mice was able to induce cartilage when brought into contact with

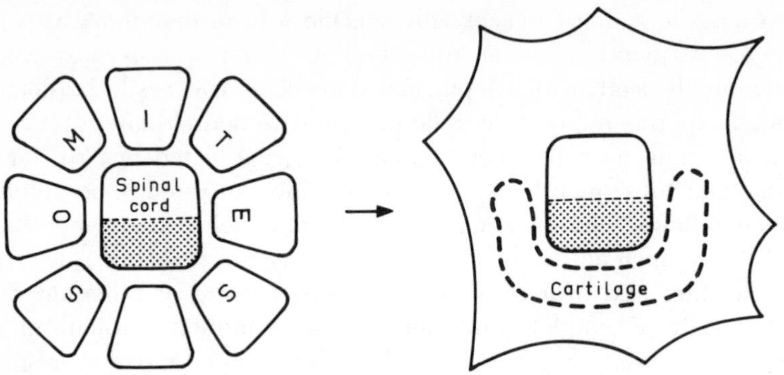

Fig. 4.12 An experimental set-up for the analysis of cartilage induction by the ventral spinal cord (dark). (After J. Lash. Am. J. Obstet. Gynecol. **90:** 1193 [1964].)

somite tissue from normal donors, but the mutant somite tissue did not respond to an inductive stimulus from normal spinal cord tissue.

Similar loss of competence, apparently combined with a simultaneous failure in the inductor tissue, has been described by Moore (1947) in hybrids of *Rana pipiens* × *Rana sylvatica.* These nonviable hybrids cease their development at the gastrula stage and survive

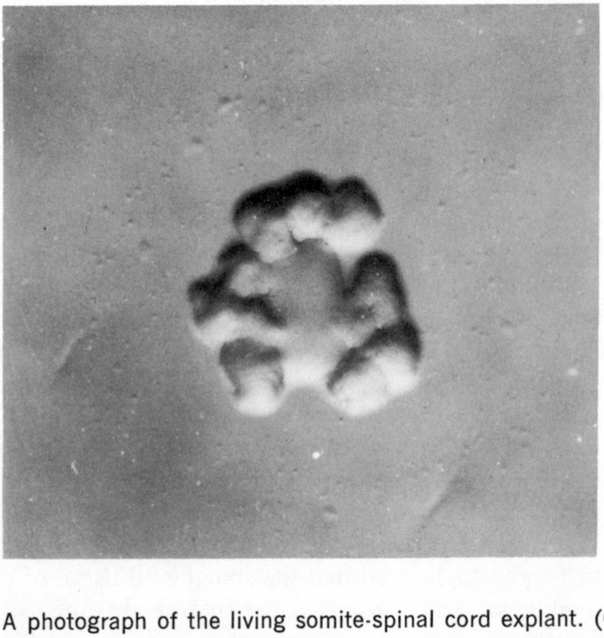

Fig. 4.13 A photograph of the living somite-spinal cord explant. (Courtesy of Dr. J. Lash.)

a few days as blocked gastrulae. The reason is apparently a failure in primary induction, and the transplantation experiments of Moore have indicated that both the competence of the ectoderm and the inductive capacity of the archenteron roof are defective.

GROWTH

In the previous section a number of examples were given in which the environment regulates the control of differentiation and in which certain congenital defects could be traced back to these mechanisms. Regarding the growth of embryonic tissues, our knowledge of the control mechanism is even more scattered and fragmentary and the molecular basis of growth control mechanisms is virtually unknown. Consequently, a discussion of the possible failures in these mechanisms as a cause of maldevelopment has to be based mainly on speculations regarding certain anatomic malformations apparently caused by abnormal growth.

A great many specific humoral factors controlling normal growth have been suggested and apparently both promotors and inhibitors circulate in the organism. The best known of these are hormones; both decreased endocrine function and overproduction of certain hormones may lead to severe growth disturbances. Humoral factors are apparently involved in the regeneration or compensatory hyperplasia of partially removed parenchymal organs (liver and kidney) and there are observations indicating that such factors can pass the maternal-fetal barrier and exert their specific effect on the corresponding fetal organs. So far, the only tissue-specific growth factor (see also below) partially isolated and purified seems to be the one stimulating erythropoiesis. Another very well-documented factor, the nerve growth-promoting factor, should be mentioned (Levi-Montalcini and Angeletti, 1965). This protein, isolated from various biologic sources, seems to have a specific growth-promoting action on cells of the sympathetic nervous system both *in vitro* and *in vivo*.

Without being able to go deeper into these and other possible mechanisms controlling the growth of embryonic cells, some examples will be given in which abnormal growth of a certain organ, tissue, or part of the embryo has led to a variety of malformations. Here, as elsewhere in teratology, it is often difficult, if not impossible, to distinguish sharply between abnormal growth and other developmental mechanisms (migration of cells, tissue interactions, tissue degeneration), as will be stressed below in several connections. Abnormal growth may be divided for practical convenience into growth inhibition (retardation), overgrowth, misplaced growth, and uncontrolled growth.

Growth Inhibition

Growth can be inhibited either at a particular time or in a particular organ, or systemic growth can be specifically inhibited over a long period. Two examples may illustrate these alternatives and their consequences.

Figure 4.14 shows three types of defects in the interatrial septal complex of the heart, defects well known from both experimental work and clinical practice. All three, apparently, result from inhibited growth and, depending on the *time* of the inhibition, the result varies. In type A the septum primum has never reached the endocardial cushion and a persistent foramen ovale primum is the result. In type B the development of the septum primum has been completed, but inhibition at a somewhat later stage of development has retarded the growth of the septum secundum, with the consequence that the foramen secundum remains open. The third type (C) represents a combination of these two. In this connection it is of interest that these malformations are common in the congenital rubella syndrome. Both experimental data and quantitative studies on children of mothers who had rubella during pregnancy indicate that the virus may inhibit cell multiplication and, hence, the frequent results of inhibited growth in the interatrial septa may be directly attributed to the virus. (Here it should be stressed that, in addition to true growth inhibition, an abnormal or excess resorption of the septa may lead to a similar anatomic end result.)

As an example of a systemic growth inhibition, congenital hydro-

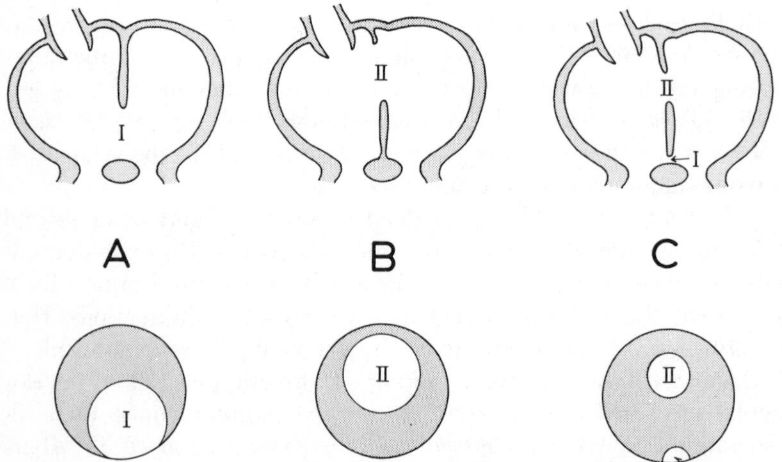

Fig. 4.14 A scheme of different types of interatrial septal defects.

cephaly in mice (*ch*) may be mentioned (Grüneberg, 1964). This rather complex syndrome, comprising defects in the chondrocranium, the vertebrae, the appendicular skeleton, and the visceral skeleton, can be traced back to a simple retardation of the growth of the blastemata in the membranous skeleton. Throughout development, the blastemata are reduced in size and the effect is not limited to a certain period of time. Because chondrification does not take place until the blastemata reach a minimum size, cartilage development is retarded and leads to the complex syndrome with secondary hydrocephalus.

Overgrowth

Increased body or organ size has been described in connection with several congenital syndromes: large children of diabetic mothers and the hyperplastic thymus associated with anencephaly may be mentioned. Systemic overgrowth seems to be the cause of a condition known as multiple exostoses, characterized by multiple cauliflower-like outgrowths from the cartilage bones. Local overgrowth as a causative mechanism for malformations, on the other hand, has received little recognition. Patten (1957) has analyzed several human embryos with malformations of the myeloschisis group (imperfect closure of the neural tube). This common type of malformation has generally been considered to result from a developmental arrest, where retarded growth has prevented fusion of the neural folds (Fig. 4.15). In the above mentioned cases, however, Patten could show a definite overgrowth of the neural material at the site of the schisis, yielding a tissue mass up to four times that on either side of the defect. A similar overgrowth has been noted in mice with a certain lethal mutation at the T-locus. In this allele, known as t^{w18}, the primitive streak grows excessively and bulges into the overlying ectoderm, so that a partially double neural tube often results (Dunn and Bennett, 1964). Myeloschisis may, in addition, serve as an example of the complexity of mechanisms of malformation. Owing to imperfect closure of the neural folds, its inductive action (page 95) is topographically altered (Fig. 4.15). Consequently, vertebral chondrogenesis takes place in an abnormal site and gives rise to an open neural canal, as indicated in the scheme.

Misplaced Growth

In numerous examples of malformations, the morphologic end result could easily be explained by growth that is normal in character but misplaced in location. Here, it will suffice to present one example

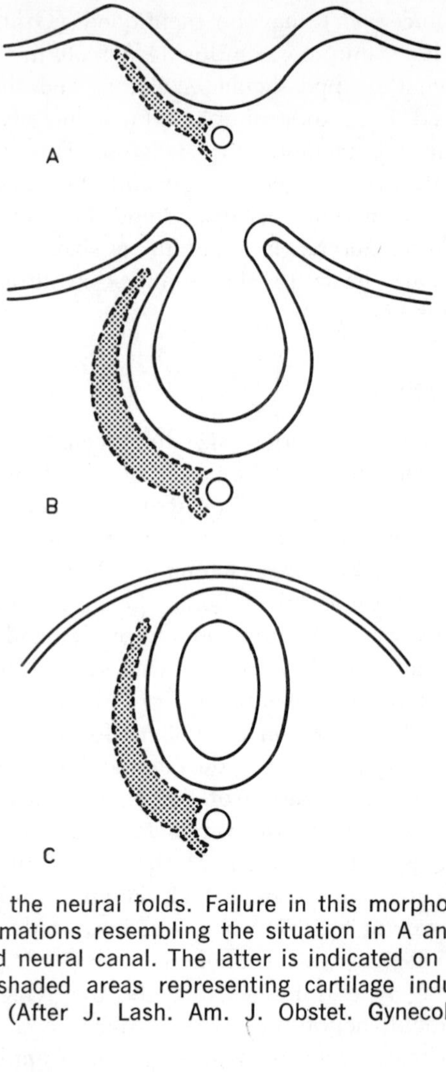

Fig. 4.15 Closure of the neural folds. Failure in this morphogenetic process leads to malformations resembling the situation in A and B with an open neural tube and neural canal. The latter is indicated on the left side of the pictures as shaded areas representing cartilage induced by the ventral spinal cord. (After J. Lash. Am. J. Obstet. Gynecol. **90:** 1193 [1964].)

in which a seemingly complex malformation can be traced back to a single event of misplaced growth. The congenital malformation of the heart known as the "tetralogy of Fallot" (Fig. 4.16) comprises the following four features: stenosis of the pulmonary artery, dextroposition of the aorta, ventricular septal defect, and hypertrophy of the right ventricle. This complex anomaly can be explained as the result of uneven partition of the truncus arteriosus, as schematized in Fig. 4.17. Uneven division of the truncus, as shown in the scheme, would give rise to a constricted pulmonary artery and an abnormally wide opening of the aorta, which overrides the interventricular sep-

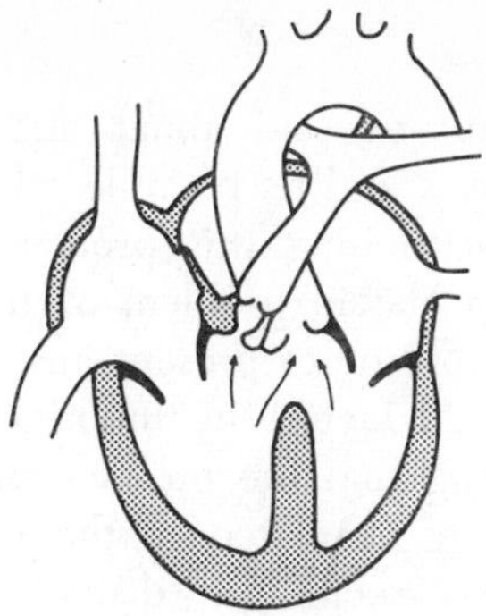

Fig. 4.16 Tetralogy of Fallot. (After J. Langman. Medical Embryology. Williams and Wilkins Co., Baltimore [1963].)

tum. Moreover, the asymmetry of the ridges would derange the normal fusion of the three components of the septum, the two ridges and the endocardial cushion of the ventricles, thus leading to a septal defect. (The hypertrophy of the right ventricle is finally a consequence of the stenotic pulmonary artery.)

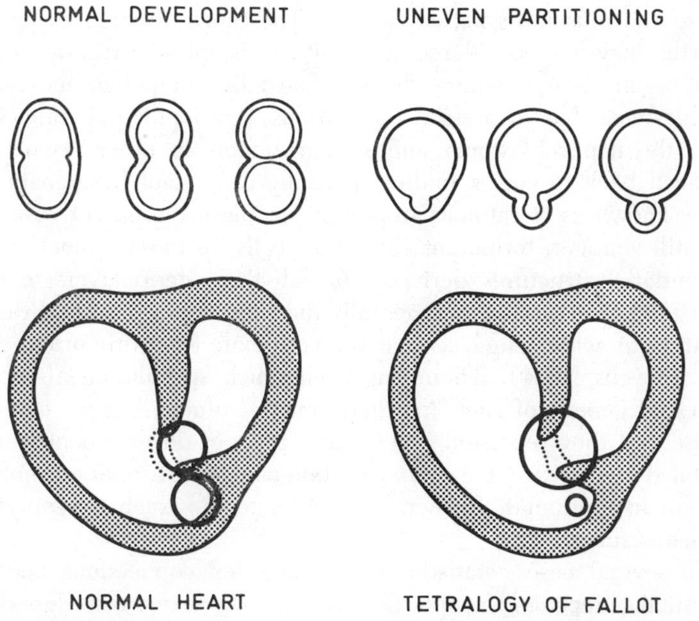

Fig. 4.17 Uneven partition of the truncus arteriorus leading to the tetralogy of Fallot. (After B. M. Patten. Human Embryology. Blakiston Co., New York [1953].)

Uncontrolled Growth

Finally, a word should be said about uncontrolled, neoplastic growth during embryogenesis and its possible relation to the problem of maldevelopment. In mentioning this process, it is certainly not our intention to bring the classic problem of the embryonic origin of tumors into the discussion, or to present any theories on the role of genetic and environmental factors in tumorigenesis. We will only present some data indicating that the problem of abnormal embryogenesis may not be entirely unrelated to the unsolved problem of cancerogenesis. These two seemingly distant problems may be brought together by three different approaches: exploration of congenital tumors, statistical study of the concurrence of cancer and congenital defects, and an analysis of the environmental factors common to teratogenesis and oncogenesis. A variety of congenital tumors have been described, ranging from misplaced, hamartomatous tissue masses to true invasive, metastasizing malignant neoplasms (see Willis, 1958). Only one example will be given here; Wilms' tumor of the kidney (nephroblastoma) consists of immature tissue components often resembling tubular or glomerular components of the embryonic kidney. In addition, other mesenchymal structures may be seen intermingled with normal kidney components (connective tissue, adipose and muscle tissue, anl so on). The tumor, sometimes present at birth, may be considered a result of neoplastic transformation of an organ anlage, where the still partially competent mesenchymal blastema has expressed its various developmental capacities. While the nephroblastoma and similar tumors in other organs can be traced back to corresponding, partially competent blastemata, the tumors known as "teratomas" represent neoplasms apparently derived from still younger, totipotent embryonic cells. In these tumors, highly differentiated structures derived from all three germ layers may be found (Fig. 4.18) and experimentally they have been produced during prenatal life, which suggests that they originate from primordial germ cells (Stevens, 1964). There has been much speculation about the origin and genesis of such "parthenogenetic" tumors but we can only surmise that they represent a very early error of development leading to total dysgenesis of the embryo. Malignant transformation, not uncommon in teratomas, furthermore serves to link such dysgenesis to true neoplastic growth.

In several cases, statistics have suggested connections between congenital defects and neoplastic diseases: where two disorders occur in the same individual more frequently than would be expected from their general incidence, a causal relationship or a common causative factor may be suspected. Regarding neoplasms and congenital defects,

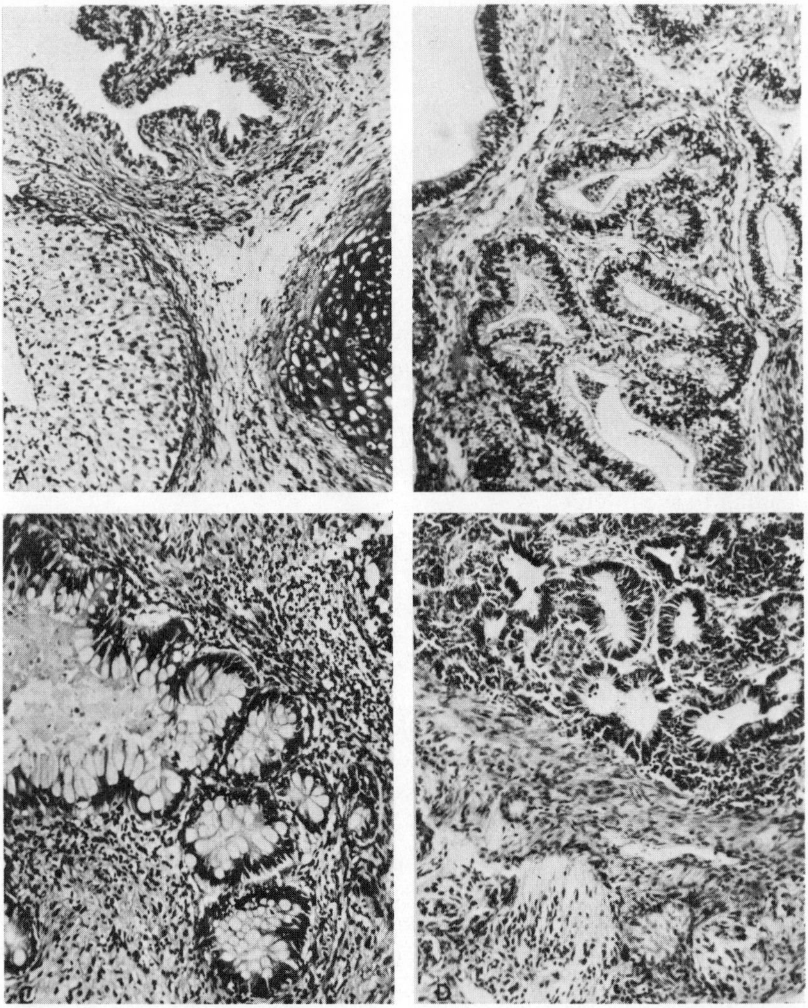

Fig. 4.18 Various highly differentiated tissue components in a human testicular teratoma: squamous epithelium (A), epithelium resembling respiratory lining (B), intestinal-type epithelium (C), masses of neutral tissue (D).

two such correlations can be mentioned: children with Down's syndrome (page 66) develop leukemia three times as frequently as others; and Wilms' tumor, described above, seems to be associated with certain congenital defects—the incidence of the defect aniridia is estimated to be 1:50,000, whereas in a (small) series of Wilms' tumors its frequency was 1:73 (Miller, 1966). Here, a common causative factor may be suspected. In a recent review, DiPaolo and Kotin (1966) have collected an extensive list of factors known to be both teratogenic and oncogenic (viruses, chemicals, hormones, radiation). These authors conclude that "a compound found to be carcinogenic to mature cells might be teratogenic to immature, embryonic cells."

DEGENERATION

Cell death and subsequent tissue necrosis are known to be direct consequences of a great many exogenous teratogens. Irradiation, viruses, and toxic chemicals frequently cause tissue loss in adults and even more so in the susceptible embryo. Consequently, developmental defects can be classified as due either to selective cell death or to generalized effects on the whole embryo. Depending on its extent, the lesion will result in loss of the embryo, defects in certain tissues or lesions than can subsequently be restored by regenerative processes. The fact that cell death and disintegration constitute a basic regulatory system in normal development deserves some comment before we proceed to a discussion of the persistent tissue damage leading to such defects.

Normal Degeneration

Normal ontogenesis is always accompanied by resorption of various tissue components and whole organs. Tails and gills of amphibian larvae, embryonic blood vessels, parts of the genitourinary tract in higher vertebrates and cartilage of the calcifying bones are normally resorbed during embryogenesis, to mention only a few examples. In addition to such resorptive processes, there are several recent studies demonstrating the morphogenetic role of focal tissue resorption and once again we may take as an example the development of the limb bud. Saunders and Fallon (1966) have extensively reviewed their studies on the topic, which demonstrate convincingly how such controlled, regular cell death plays a central role in the carving out and shaping of the wing (and limb). If such an area, destined to necrotize, is transplanted from its original site to somite tissue at an early stage of development, cell death is not prevented and the tissue disintegrates in its new location (controls grafted from other parts of the wing survive the grafting). This indicates that the cells have really been predetermined to die at a certain stage of development, independent of their environment at the time of execution. Saunders speaks of the "death clock" in these cells and Lash (1964) refers to it as "inherent obsolescence." The factors underlying such determined degeneration are not known, but one example illustrating the role of tissue interaction should be given here. The separation of digits in the chick limb bud involves definite degeneration of the interdigital tissue. This process is less marked in the hind limb of the duck, where webbing persists. In his combination experiments, described on page 91, Zwilling (1956) has made reciprocal combinations of chick and duck limb bud components (mesoblast plus ectodermal ridge). When the chick mesoblast was covered by ectoderm of duck hind limb, the latter apparently inhibited the degeneration of the interdigital tissue and webbing resulted.

After these comments on normal degeneration, we must now discuss the disturbances in this process that lead to maldevelopment. In doing so, there is every reason to follow Zwilling's (1955) classification into abnormal and excessive degeneration and failure to degenerate.

Abnormal Degeneration

In accordance with the introductory remarks, "abnormal" degeneration refers to tissue loss at sites and stages where this does not occur during normal ontogenesis. The direct effects of exogenous teratogens mentioned above fall into this category and there is no need to list a great number of examples here. However, we may refer to the effect of nitrogen mustard on the limb bud mesoderm (page 92), the effect of x-rays on the developing central nervous system (page 171), and the cytopathic effect of viruses (page 211). The actual problem seems to be the selective nature of this effect. Why are some cells more vulnerable than others and succumb to a treatment which leaves others apparently intact? Some of the rather scanty information relating to this problem will be dealt with in Chapter 5, "Sensitive Periods in Development."

Abnormal resorption has been shown to occur in several genetic defects where differentiation seems to proceed normally up to a certain point, but is followed by localized tissue necrosis. A recessive mutant in rabbits causes brachydactyly in homozygous animals and the first change seems to be a progressive necrosis in the limb bud followed by healing of the wound and abnormal limbs (Inman, 1941).

In a whole group of mutants suffering from the walzer-shaker type syndrome in the mouse, the defect is to be found in the inner ear. The labyrinth appears quite normal at the time of birth and continues to develop normally for several days. Degenerative changes then occur in the organ of Corti and in the spiral ganglia and there is ultimately complete necrosis (Grüneberg, 1956). Finally, the T-locus mutants should be mentioned again. In the lethal homozygote t^{w1}/t^{w1}, cell degeneration and death are visible in the ventral neural tube from the eighth day on. This progressive degeneration may lead to total destruction of the entire ventral neural tube, but in some embryos this is followed by reparative processes. Restoration of the lost tissue results, but the regenerate lacks normal tissue architecture. This restoration process is, moreover, of some theoretic interest since it shows that repair of minor degenerative lesions is possible during subsequent development.

Excessive Normal Degeneration

A second possible mechanism related to tissue resorption is the abnormal spread of the normal necrotic foci (see above). As Zwilling (1955) has said, it is probable that these foci are normally kept under control by inhibitor factors and hence lack of these hypothetic factors or their elimination may lead to abnormal spread of disintegration. The following two examples appear to illustrate such situations. In normal development, the future knee joint is delineated by a zone of dying cells. Administration of insulin to chick embryos is known to cause micromelia and Zwilling (1959) has shown that the treatment results in a spread of the normal degenerating zone. If bone rudiments of such insulin-treated embryos are grown in organ culture, no such effects are noted, so that the *in vivo* effect is best explained by assuming that insulin interferes with the normal control mechanism restricting tissue resorption.

The same author has analyzed a rumpless mutant in the chick, in which the entire caudal part of the vertebral column is missing. In normal chick embryos, one of the determined zones of cell death is to be found posterior to the undifferentiated tail bud. Examination of the mutant embryos revealed that this focus was more extensive than it was in normal embryos and that the extent of the degenerating tissue corresponded to the degree of reduction of the caudal vertebrae.

Both these examples seem to demonstrate the potential danger of the normal "death clock," but very little is known about the effect of different exogenous factors that might affect this system, either by interfering with the timing of the degeneration or with the mechanisms controlling its spread.

Failure to Degenerate

A great variety of developmental vestiges representing remnants of tissue components or whole organs that normally degenerate are known to the pathologist: persistent ductus Botalli, cysts derived from the thyroglossal duct, remnants of Rathke's pouch, Meckel's diverticulum, and vestiges of the urogenital duct represent such failures of normal degeneration. (For a comprehensive presentation, the reader is referred to Willis' book, 1958.)

We may ask whether such failures of normal degeneration of

cells and tissues may lead to true malformations in the restricted sense. The answer is in the affirmative as illustrated in a recent analysis of reopening of the esophagus in a certain mutant chick (Allenspach, 1966). In normal embryos, the esophagus becomes completely occluded at stage 26 (5 days) and its subsequent reopening is completed 2 to 3 days later. This is chiefly the consequence of desquamation of the esophagal epithelium leading to a recanalization of the lumen. In the crooked neck dwarf mutant, the esophagus is still largely occluded beyond day 12 and it has been suggested that the embryo suffers from starvation because it is unable to swallow the nutritious amniotic fluid. Analysis of the course of development preceding this stage showed that the process was normal up to stage 33 (7 to 8 days), but that the degeneration of the epithelium was thereafter incomplete and resulted in persisting cellular strands, causing occlusion of the lumen.

Finally, the development of the limb bud should be referred to again. We have already described the inductive interactions during this event (page 90) and stressed the importance of normal degeneration of the mesodermal component (page 104). A recent study by Hinchliffe and Ede (1967) deals with limb bud development in a mutant fowl strain and shows how failure to degenerate may lead to subsequent abnormalities in the interactive processes and finally to a structural defect. The talpid^3 mutant analyzed is characterized by abnormally shaped legs with a greatly increased number of digits. Analysis indicated that the normally occurring cell death seen as massive zones of necrosis in the mesenchymal part of the bud was completely lacking in the mutant strain (Fig. 4.19). Thus, the abnormal shape of the limb bud depends on the absence of the carving effect of necrosis. Furthermore, the mesoblast remains larger than in normal embryos and, consequently, the area providing the maintenance factor for the ectoderm (page 90) is not limited by necrotic zones. As a result, the ectodermal ridge continues to grow and presses an abnormally wide inductive template on the mesoblast, leading to polydactyly. The results are thus in agreement with the classic observations by Zwilling (page 91), showing the decisive role of the mesoblast in the development of polydactyly, except that they seem to locate the primary failure one step earlier to a defect in the "death clock" of the cells in one of the interacting components.

In the recombination experiments already described, ducklike webbing of the foot was obtained when chick limb bud mesoderm was covered by duck ectoderm, and we concluded that the latter may have an inhibitory effect on the normal morphogenetic degeneration of the interdigital tissue. Very similar syndactyly of the soft parts of the chick limb has been produced by an exogenous teratogen,

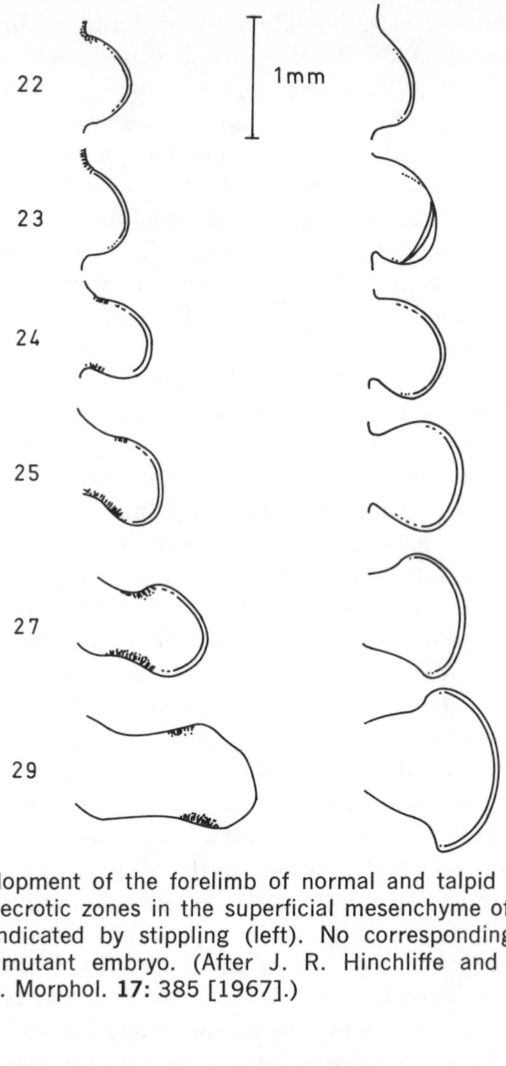

Fig. 4.19 Development of the forelimb of normal and talpid mutant fowl embryos. The necrotic zones in the superficial mesenchyme of the normal limb-bud are indicated by stippling (left). No corresponding zones are visible in the mutant embryo. (After J. R. Hinchliffe and D. A. Ede: J. Embryol. Exp. Morphol. **17:** 385 [1967].)

Janus green, by treating the embryos prior to the stage of interdigital necrosis (Menkes and Delanu, 1964; Saunders and Fallon, 1966). In both experimental series, the interdigital tissue failed to degenerate, and consequently the defect may represent another example of maldevelopment resulting from inhibition of normal degeneration. On the other hand, however, as the latter authors stress, the failure to degenerate might not represent the primary effect of the teratogen, but might be a consequence of disturbances in the preceding developmental events.

SUGGESTED READINGS

Review Articles

De Haan, R. L. Development of form in the embryonic heart. An experimental approach. Circulation. **35**: 821–833 (1967).

DiPaolo, J. A. and P. Kotin. Teratogenesis-oncogenesis: A study of possible relationships. Arch. Pathol. **81**: 3–23 (1966).

Ebert, J. D. Interacting Systems in Development. Holt, Rinehart and Winston, New York (1965).

Fraser, F. C. Some genetic aspects of teratology. *In* Teratology, Principles and Techniques, (J. G. Wilson and J. Warkany, eds.), pp. 21–38. The University of Chicago Press, Chicago (1965).

Grobstein, C. Inductive tissue interaction in development. Adv. Cancer Res. **4**: 187–236 (1956).

Grüneberg, H. The Pathology of Development. A Study of Inherited Skeletal Disorders in Animals. Blackwell, Oxford (1963).

———. The genesis of skeletal abnormalities. *In* Congenital Malformations, (M. Fishbein, ed.), pp. 219–223. The International Medical Congress, Ltd., New York (1964).

Gustafson, T. and L. Wolpert. The cellular basis of morphogenesis and sea urchin development. Intern. Rev. Cytol. **15**: 139–214 (1963).

Holtfreter, J. and V. Hamburger. Embryogenesis: Progressive differentiation. *In* Analysis of Development, (B. H. Willier, P. A. Weiss and V. Hamburger, eds.), pp. 230–296. W. B. Saunders Co., Philadelphia (1955).

Holtzer, H. The induction and maintenance of vertebral cartilages. *In* Congenital Malformations, (M. Fishbein, ed.) pp. 233–239. The International Medical Congress, Ltd., New York (1964).

Kalter, H. Interplay of intrinsic and extrinsic factors. *In* Teratology, Principles and Techniques. (J. G. Wilson and J. Warkany, eds.), pp. 57–80. The University of Chicago Press, Chicago (1965).

Lash, J. W. Tissue interaction and specific metabolic responses. Chondrogenic induction and differentiation. *In* Cytodifferentiation and Macromolecular Synthesis, (M. Locke, ed.), pp. 235–260. Academic Press. New York (1963).

Miller, R. W. Relation between cancer and congenital defects in man. New Engl. J. Med. **275**: 87–93 (1966).

Patten, B. M. Varying developmental mechanisms in teratology. *In* Congenital Malformations. (J. Warkany, ed.), Pediatrics, Suppl. **19**: 734–748 (1957).

Runner, M. N. General mechanisms of teratogenesis. *In* Teratology, Principles and Techniques, (J. G. Wilson and J. Warkany, eds.), pp. 95–103. The University of Chicago Press, Chicago (1965).

Saunders, J. W., Jr. and J. F. Fallon. Cell death in morphogenesis. *In* Major Problems in Developmental Biology, (M. Locke, ed.), pp. 289–314. Academic Press, New York (1966).

Saxén, L. and S. Toivonen. Primary Embryonic Induction. Academic Press, London (1962).

Trinkaus, J. P. Mechanisms of morphogenetic movements, *In* Organogenesis, (R. L. DeHaan and H. Ursprung, eds.), pp. 55–104. Holt, Rinehart and Winston, New York (1965).

Waddington, C. Developmental mechanisms. Introduction. *In* Congenital Malformations, (M. Fishbein, ed.), pp. 213–218. The International Medical Congress, Ltd., New York (1964).

Willis, R. A. The Borderland of Embryology and Pathology. Butterworth and Co., London (1958).

Wolff, Et. The experimental production and the explanation of phocomelia in the chick embryo. *In* International Workshop Teratol. pp. 84–94. Copenhagen (1966).

Zwilling, E. Teratogenesis. *In* "Analysis of Development," (B. H. Willier, P. A. Weiss and V. Hamburger, eds.), pp. 699–719. W. B. Saunders Co., Philadelphia (1955).

Special Articles

Allenspach, A. L. The reopening process of the esophagus in the normal chick and the crooked neck dwarf mutant. J. Embryol. Exp. Morphol. **15**: 67–76 (1966).

Chase, H. B. and E. B. Chase. Studies on an anophthalmic strain of mice. I. Embryology of the eye region. J. Morphol. **68**: 279–301 (1941).

DeHaan, R. L. Oriented cell movements in embryogenesis. *In* Biological Organization at Cellular and Supercellular Level. (R. J. C. Harris, ed.), pp. 147–165. Academic Press, New York and London (1963).

———. Cell interactions and oriented movements during development. J. Exp. Zool. **157**: 127–138 (1964).

Dunn, L. C. and D. Bennett. Abnormalities associated with a chromosome region in the mouse. I. Transmission and population genetics of the t-region. II. Embryological effects of lethal alleles in the t-region. Science **144**: 260–267 (1964).

Gluecksohn-Waelsch, S. and T. R. Rota. Development in organ tissue culture of kidney rudiments from mutant mouse embryos. Develop. Biol. **7**: 432–444 (1963).

Grüneberg, H. Hereditary lesions of the labyrinth in the mouse. Brit. Med. J. **12**: 153–157 (1956).

Inman, O. R. Embryology of hereditary brachydactyly in the rabbit. Anat. Record **79**: 483–505 (1941).

Lahti, A. and L. Saxén. Effect of hydrocortisone on the closure of palatal shelves *in vivo* and *in vitro*. Nature **216**: 1217–1218 (1967).

Levi-Montalcini, R. and P. U. Angeletti. The action of nerve growth factor on sensory and sympathetic cells, *In* Organogenesis, (R. L. DeHaan and H. Ursprung, eds.), pp. 187–198. Holt, Rinehart and Winston, New York (1965).

Menkes, B. and M. Deleanu. Leg differentiation and experimental syndactyly in chick embryo. II. Experimental syndactyly in chick embryo. Rev. Roum. Embr. Cyt. **1**: 69–77 (1964).

Moore, J. A. Studies in the development of frog hybrids. II. Competence of the gastrula ectoderm of Rana pipiens ♀ x Rana sylvatica ♂ hybrids. J. Exp. Zool. **105**: 349–370 (1947).

Saunders, J. W., Jr. The proximo-distal sequence of origin of the parts of the chick wing and the role of the ectoderm. J. Exp. Zool. **108**: 363–404 (1948).

Stevens, L. C. Experimental production of testicular teratomas in mice. Proc. Nat. Acad. Sci. **52**: 654–661 (1964).

Zwilling, E. Interaction between limb bud ectoderm and mesoderm in the chick embryo. II. Experimental limb duplication. III. Experiments with polydactylous limbs. IV. Experiments with a wingless mutant. J. Exp. Zool. **132**: 173–187, 219–239, 241–253 (1956).

Zwilling, E. Micromelia as a direct effect of insulin—Evidence from *in vitro* and *in vivo* experiment. J. Morphol. **104**: 159–179 (1959).

5

Sensitive Periods in Development

In the previous chapter we briefly mentioned some of the normal control mechanisms that guide differentiation and ensure synchronized development: morphogenetic tissue interactions, regulated migrations of cells and cell colonies, controlled proliferation, cell death, and so on. Under such control systems, embryonic development is relatively well canalized and buffered against abnormal environmental conditions. On the other hand, these systems leading to reorganization and sequential differentiation of the embryonic tissue seem to create the most vulnerable phases of embryogenesis, the epigenetic crises (Waddington, 1966). Two evident reasons for this causal relationship between the control mechanisms and the sensitive periods may be mentioned. First, the control of differentiation by the micro-environment of the cells seems to be a highly vulnerable process, the cellular contacts being easily interfered with and the chemical factors presumably involved being destructible. Second, by definition, these environmental triggers stimulate new morphogenetic processes characterized by increased synthetic activities of the responding cells and there is good reason to believe that such active stages of differentiation are more sensitive to disturbance than resting phases. In the following paragraphs, the role of such critical periods or sensitive stages of development will be analyzed and examples of several epigenetic crises will be given, subsequent to the presentation of some of the classic concepts in this field.

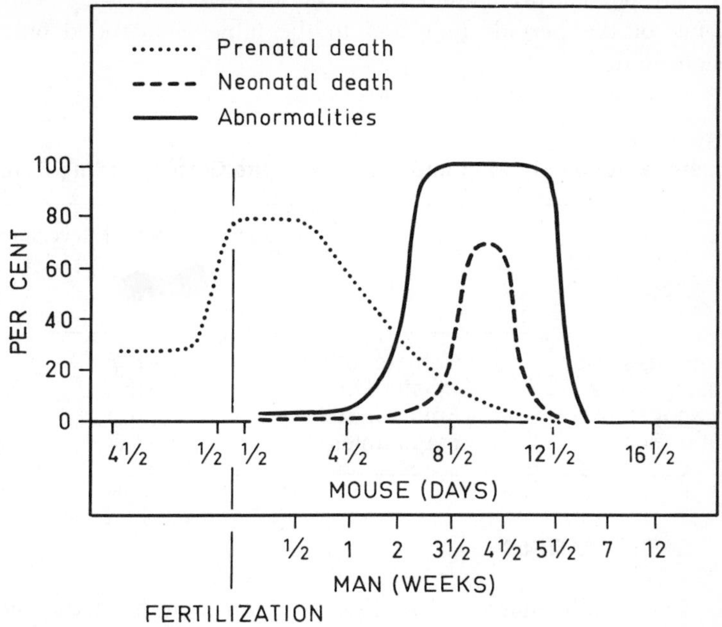

Fig. 5.1 The effect of irradiation at different stages of development on the survival and maldevelopment of the embryos. Original experiments performed on mice and the corresponding periods for man have been calculated. (After L. B. Russell and W. L. Russell. J. Cell. Comp. Physiol. **43:** Suppl. 103 [1954], and A. B. Brill and E. H. Forgotson. Am. J. Obstetr. Gyneol. **90:** 1149 [1964].)

The classic experiment demonstrating the sensitive periods during embryogenesis is illustrated in Fig. 5.1. Irradiation of the pregnant mouse with 200 *r* during different stages of gestation has a clearcut effect which can be divided into three distinct periods. During early development, corresponding to gametogenesis and blastogenesis, a high percentage of the embryos will be killed, but the survivors undergo normal development. During the second period, the time of major organogenesis, the malformation rate may be as much as 100 percent, but the embryos survive, at least through the intrauterine period. After this sensitive period, when organogenesis is more or less completed, the embryos seem to be protected against exogenous teratogens and develop normally.

Although this basic scheme may not be fully applicable to congenital defects in a wider sense (as will be shown later), we may use it here for convenience and consequently divide embryogenesis into four periods, each of which responds in its own characteristic way to harmful exogenous influences (Table 5.1). The periods naturally are not sharply limited and certainly partially overlap. The exact timing of the periods indicated in the table is intended only as a rough guide.

TABLE 5.1
Sensitive Periods in Man and Mouse and the Corresponding Defects.

		Days of Development	
		Man	*Mouse*
Gametogenesis	Gametopathies	–1	–1
Blastogenesis	Blastopathies	1–15	1–6
Organogenesis	Embryopathies	16–72	7–12
Maturation	Fetopathies	73–280	13–21

GAMETOPATHIES

The results illustrated in Fig. 5.1 show that irradiation is hazardous even when exposure occurs before fertilization, because damage inflicted during gametogenesis may subsequently lead to pregnancy wastage. In the previous chapter, we listed some additional environmental factors which induce defects during this period, seen as chromosomal abnormalities, and in Chapter 7 we will discuss the effect of irradiation in greater detail. Summarizing, we may state that gametogenesis of both male and female germ cells comprises a certain sensitive period during which exogenous factors may give rise to defects, either genetic or nongenetic, and consequent fetal wastage. Estimates of the magnitude of pregnancy wastage caused by abnormal gametogenesis vary greatly, but the proportion of fertilized eggs that perish before birth might be about 20 percent and there is good reason to believe that a high proportion of these can be attributed to gametopathies.

In addition to irradiation, other teratogens, like viruses and drugs, seem to act as causative agents for teratogenesis during the prefertilization period. We will later present some preliminary data suggesting that maternal virus infections may be linked with chromosomal aberrations (page 226) and will refer to virus-induced avian leusoses, where the ovum is infected during gametogenesis and gives rise to a diseased bird. Regarding drugs, thalidomide may again serve as an example. The drug has been found in rabbit semen for considerable periods after oral administration of thalidomide to the animals and some deleterious effects seen as fetal wastage have been noted in the offspring of treated male rabbits (Lutwak-Mann, 1964; Lutwak-Mann *et al.*, 1967).

Overripeness

An interesting, although still relatively obscure, risk for fetal development consists of delayed ovulation and fertilization. Already Witschi (1952) has produced malformed embryos in amphibians by de-

Fig. 5.2 Abnormal blastocysts resulting from fertilization of rat ova after experimentally induced delayed ovulation. (A) Normal blastocyst, (B) Abnormal blastocyst where only one blastomere of the 2-cell stage has developed, (C) Blastocysts composed of abnormal large cells, (D) Blastocyst apparently developed from 2 blastomeres present at the 4-cell stage. (From N. W. Fugo and R. L. Butcher. Fert. and Ster. **17:** 804 [1966].)

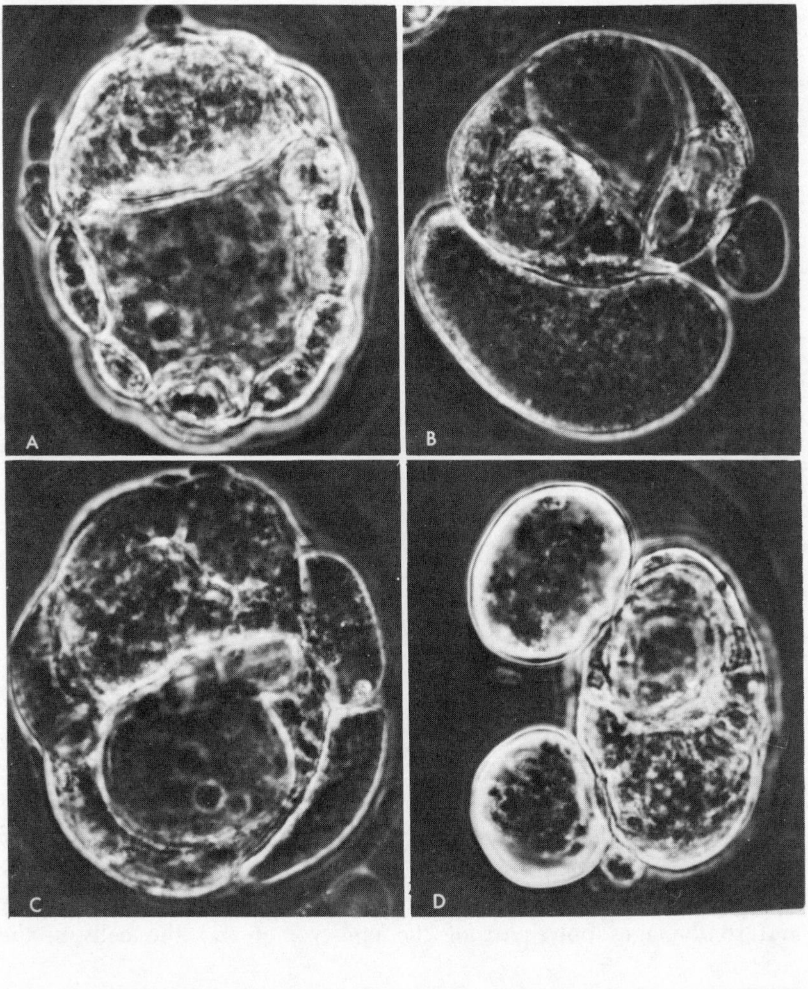

laying fertilization of the ova and more recently similar observations have been reported in higher vertebrates. The maximum period for which mammalian ova remain capable of normal development is limited and is of the order of 10 to 20 hours after ovulation. After this period of "competence," the ova gradually lose their normal developmental capacities and subsequent fertilization leads to defective or inviable zygotes (Braden, 1959; Fugo and Butcher, 1966). In addition to the gross defects illustrated in Fig. 5.2 and listed in Table 5.2, chromosomal anomalies have recently been described in embryos of rats whose ovulation had been experimentally delayed (Butcher and Fugo, 1967).

The mechanism of this gradual loss of developmental capacities, overripeness, remains to be clarified as well as its practical significance. We do not know whether this phenomenon plays a role in the early loss of human embryos or whether certain hormone treatments affecting human ovulation should be considered teratogenic because of their delaying effect. Certain recent observations on human material suggest that ova shed before or on day 14 developed normally, whereas a great proportion (12 out of 21) of those shed later were defective (Hertig, 1967). We know, moreover, that chromosomal aberrations are frequent in aborted human embryos. In view of the above observations on such abnormalities in experimentally induced overripe eggs, this mechanism has to be considered as one possible causative factor in human chromosomal aberrations, discussed in Chapter 3.

TABLE **5.2**
The Effect of Delayed Ovulation in Fertilized Rat Ova, Indicated by the Percentage of Abnormal Embryos at Different Stages of Development.[a]

	Pronuclear Stage		2-Cell Stage		Blastocyst Stage	
	Delayed	*Control*	*Delayed*	*Control*	*Delayed*	*Control*
Normal	55.3	90.0	74.7	92.3	46.4	79.7
Infertile	22.1	6.6	13.0	3.5		
Abnormal	22.6	3.4	12.3	4.2	53.6	20.3

[a] From N. W. Fugo and R. L. Butcher: Fert. and Ster. **17:** 804–814, 1966.

BLASTOPATHIES

A localized defect in an embryo, a malformation, can be induced by one of two mechanisms: (1) either all cells are alike and equally susceptible to the teratogen, but the latter exerts its action locally and so damages only part of the embryo; or (2) the cells in the

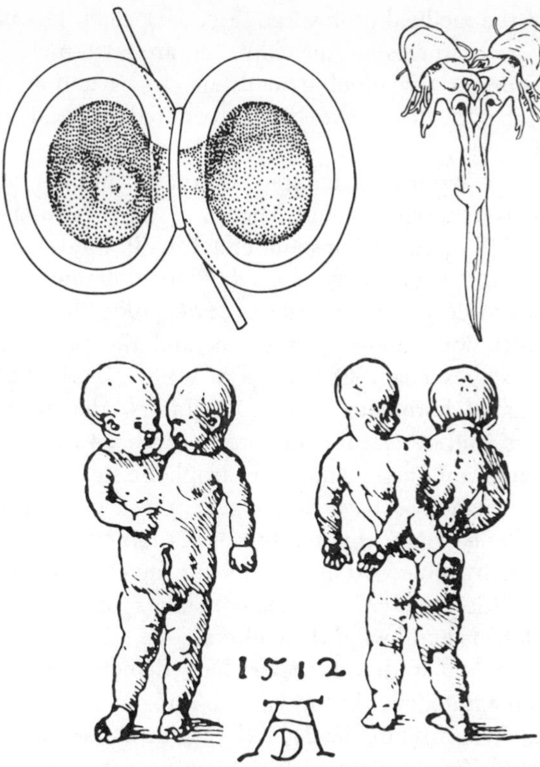

Fig. 5.3 Production of a double monster by partial separation of the cells at the 2-cell stage in an amphibian embryo and illustration of a corresponding human malformation by Dürer. (From L. Saxén and S. Toivonen. Primary Embryonic Induction, Academic Press [1962] and O. Mangold. Acta Gen. Med. Gemel. **10:** 1 [1961].)

embryo have already differentiated and developed a differential susceptibility to a teratogen acting on the whole embryo (irradiation, metabolic inhibitors), and consequently affecting only certain cells. In the earliest stages of development, the evidence indicates that the cells are still undetermined and alike: a two-cell stage can be separated with a hair loop into two parts, each of which will develop into a complete embryo, indicating that at this stage the cells are totipotent. Further, if one of the cells is destroyed, the remaining half will develop into a normal embryo, showing the capacity to "replace" the destroyed half. On the other hand, malformations of certain types can already be produced at this stage by local manipulation, such as incomplete separation of the two cells with consequent derangement of their mutual interaction. The result here would be a double monster well known to both experimental embryologists and members of the medical profession (Fig. 5.3). Such lesions arising from the very early stages of development are extremely rare, although in addition to true double monsters, twins and even sacral teratomas (sacral parasites) have been attributed to local lesions in the early zygote.

As long as we regard an embryo as totally undifferentiated, a general exogenous teratogen would lead to a similar response in all cells and result either in a transient effect with total recovery or in destruction of the whole zygote. Such general teratogens do act upon early stages of development. Adams *et al.* (1961) have employed the flatmount technique allowing a direct and detailed analysis of early embryonic stages (page 26). A large series of different teratogens tested included hormones, mitotic inhibitors, alkylating agents, metabolic analogs, and others; the authors describe a variety of effects on the preimplantation stages and early trophoblasts. In most cases, the consequence seemed to be a rather generalized effect and severe damage of the whole zygote apparently leading to nonviable embryos. In addition to such induced damage, genetic defects and maternal-fetal incompatibilities may also become manifest during this stage as generalized defects and loss of the embryo.

Some experimental studies, however, have demonstrated localized effects of general teratogens during early blastogenesis. Both irradiation (Rugh and Wohlfromm, 1962) and certain metabolic inhibitors (Gottschewski and Zimmermann, 1963) have been suggested to produce local defects in surviving embryos when administered soon after fertilization and well before any morphologic differentiation can be detected. Two alternative explanations seem plausible: (1) either a certain mosaicism at the metabolic level has already taken place, creating differential susceptibility of the cells; or (2) a delayed effect may be responsible for the result. For example, an indirect effect through the maternal organism or the placental tissue may be postulated.

EMBRYOPATHIES

According to the original experiments illustrated in Fig. 5.1, the sensitive period leading to true structural malformations coincides with the major organogenetic period. The result of an exogenous teratogen is no longer loss of the embryo, but rather a manifest developmental defect. A great number of experimental data obtained with other teratogens (drugs, viruses, physical factors, and so on) seem to corroborate this general rule and two such examples are illustrated in Figs. 5.4 and 5.5.

Figure 5.4 is based on recent experiments employing the potent

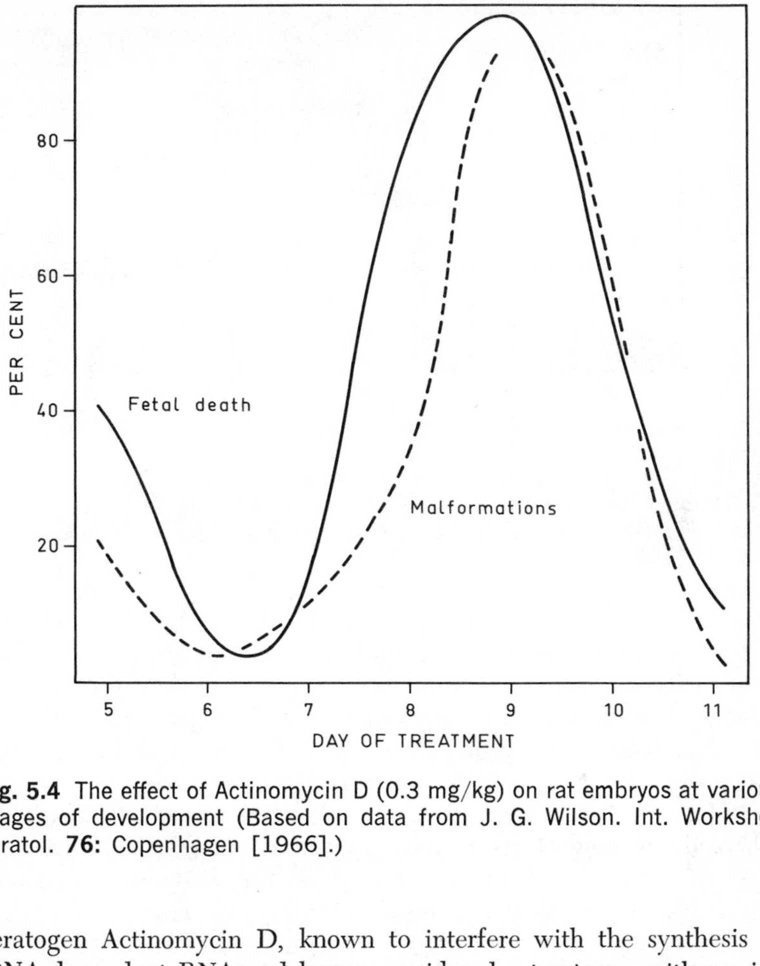

Fig. 5.4 The effect of Actinomycin D (0.3 mg/kg) on rat embryos at various stages of development (Based on data from J. G. Wilson. Int. Workshop teratol. **76:** Copenhagen [1966].)

teratogen Actinomycin D, known to interfere with the synthesis of DNA-dependent RNA and hence considered a teratogen with a wide range of effects. Here, too, true congenital defects were noted prior to the sensitive period in its restricted sense and it is interesting that this early period of blastopathies was followed by a short period of relative resistance before the onset of organogenesis. During this stage, malformations and fetal loss were noted in a high percentage, whereas by day 11 the embryo already seemed to be protected from the drug. In contrast to actinomycin, where the application and rapid effect can easily be timed, another transitory teratogenic condition is illustrated in Fig. 5.5. It has been shown that a diet deficient in folic acid (pteroylglutamic acid or PGA) causes fetal resorption and malformation when pregnant rats are kept on this diet throughout gestation. Nelson (1957) restricted this diet to 2- to 10-day periods at different stages of pregnancy and by exploring the malformations in these partially overlapping series, was able to conclude that here

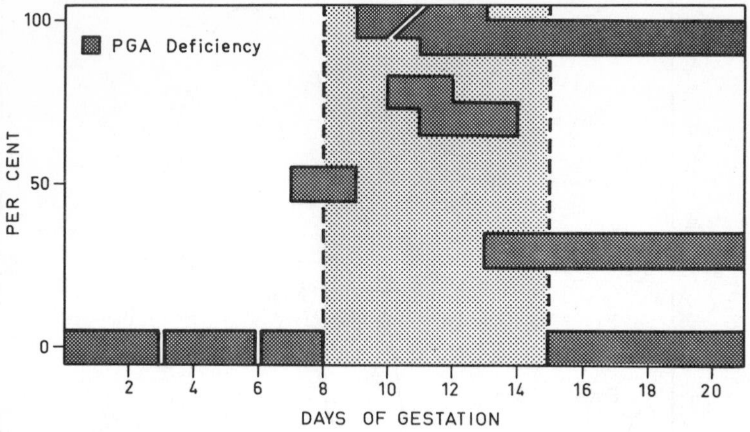

Fig. 5.5 The effect of transitory maternal folic acid deficiency on the development of rat embryos. The shaded areas indicate the period of deficiency. (Based on data from M. M. Nelson. Pediatrics, **19:** Suppl. 764 [1957].)

too, the sensitive period corresponded to the period of major organogenesis. A third example of experiments demonstrating the sensitive periods has already been illustrated in Fig. 2.8. Here the effect of vitamin A deficiency at different stages of development was demonstrated indirectly by starting treatment with this vitamin at different stages of pregnancy to rats previously kept on a deficient diet.

Next, we should mention an example based on clinical experience, where the method of overlapping periods mentioned above has been used to estimate the sensitive period of thalidomide. Accordingly, information on the consumption of this drug was retrospectively obtained from mothers of children showing limb malformations and by combining these data, the sensitive period for thalidomide embryopathy was estimated to be from days 37 through 50 (Fig. 8.11).

In conclusion, the experimental data, combined with clinical experiences, indicate that most of the epigenetic crises at which exposure to environmental teratogens is likely to lead to congenital defects and which are thus the vulnerable points of embryogenesis, fall within the period of major organogenesis. The crucial importance of the time of exposure led embryologists at one time to conclude that the timing determined the type of defect produced, irrespective of the teratogen employed. In other words, it was principally the epigenetic crisis occurring at the time of exposure that determined the consequence. This view can no longer be accepted as having general validity, but there are nonetheless definite correlations between the time of treatment and the type of defect, as will be shown in the next section.

Organ Specificity of the Sensitive Periods

Two examples have been chosen to illustrate stage-specific defects during treatment or exposure to the same teratogen (Figs. 5.6 and 5.7). One of these is based on experimental work and the other has been taken from clinical data on rubella-infected mothers.

The very definite timetable for defects of the axial skeleton is illustrated in Fig. 5.6 where the effect of short-term maternal hypoxia is analyzed. The temporal cranio-caudal gradient correlated well to the most active period of somite formation, when compared to normal development. The second example has been taken from clinical experience and indicates the sensitive periods for three common defects

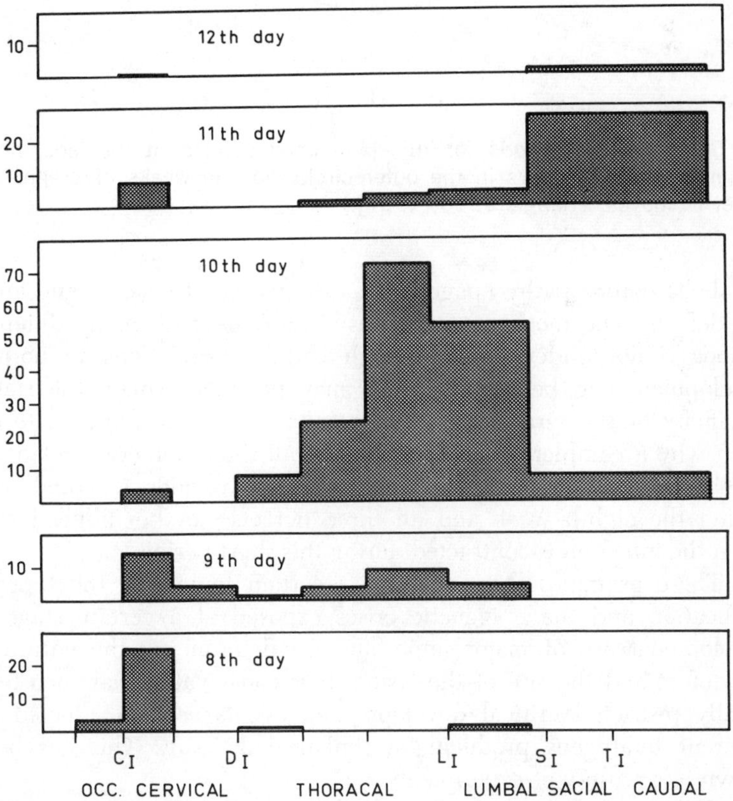

Fig. 5.6 The teratogenic effect of short-term maternal hypoxia at different stages of development on the axial skeleton. Symbols: CI, first cervical vertebra; D, thoracic; L, lumbar; S sacral; and T tail. (After U. Murakami and Y. Kameyama. J. Embryol. Exp. Morphol. **11:** 107 [1963].)

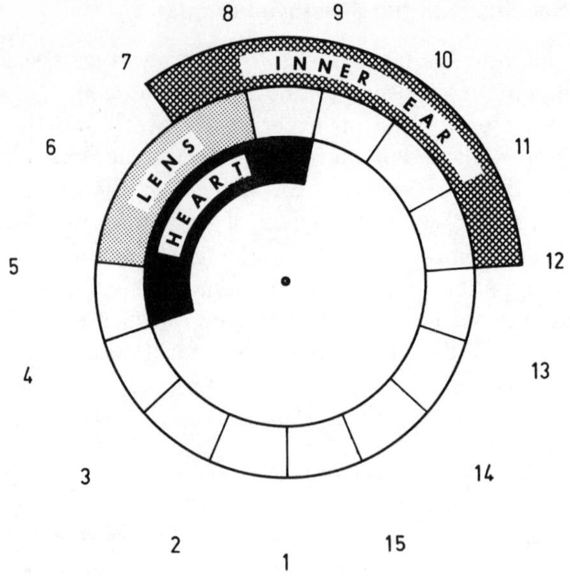

Fif. 5.7 Sensitive periods for rubella-induced defects in the lens, heart and inner ear. Numbers in the outer circle indicate weeks of pregnancy. (After B. Mayes. Triangle **3:** 10 [1957].)

in rubella embryopathy (page 219): cataract, heart disease, and inner ear defects. The mechanism responsible for these distinctly different periods is not understood, although certain correlations to normal development can be found, which may possibly explain the stage-specificity of the viral action. The results appear, in addition, to explain why a complete triad consisting of all three defects is relatively uncommon: an overlapping of the sensitive periods is noted only during the eighth week and all three defects can be induced only when the infection is contracted during this short period.

These examples show that for a certain teratogen, the time of application and the epigenetic crises exposed at a certain stage of development are of major importance in determining the nature of the defect and the site of the lesion. It is conceivable that such temporally restricted critical developmental events can be affected by different teratogens producing a similar end result. This has been shown in a variety of experiments.

Closure of the neural tube can be prevented during its sensitive period by very many different exogenous factors ranging from irradiation to organ-specific antisera, and in the list compiled by Dagg (1966) some 20 different treatments leading to cleft palate are given. But even so, it is certainly not justifiable to generalize this principle and conclude that the site of the teratogenic action and the consequent defects are entirely determined by the stage of development.

A single example will suffice to demonstrate the fallacy of this generalization: Dimethylsulphoxide (DMSO) administered to pregnant hamsters on day 8 leads to typical malformations of the head region (exencephaly), whereas injection of lead salts on the same day and under similar conditions selectively damages the caudal region (Ferm, 1967). No overlapping of these defects has been seen and the example thus indicates the specificity of the teratogen(s) versus the decisive role of the time of treatment.

Finally, we may quote an example showing how the very same defect can be induced at different times by different environmental factors. In other words, the sensitive period of a certain defect is dependent on the agent employed rather than on the developmental stage at which it is applied. Hemoglobin synthesis in early chick embryos is sensitive to a variety of metabolic inhibitors acting at different steps in the process of transcription and translation finally leading to the inhibition of specific protein synthesis. Apparently owing to differences in their points of action in this sequential process of gene-controlled synthetic activity, the sensitive periods for these different inhibitors vary greatly, as shown in Fig. 5.8.

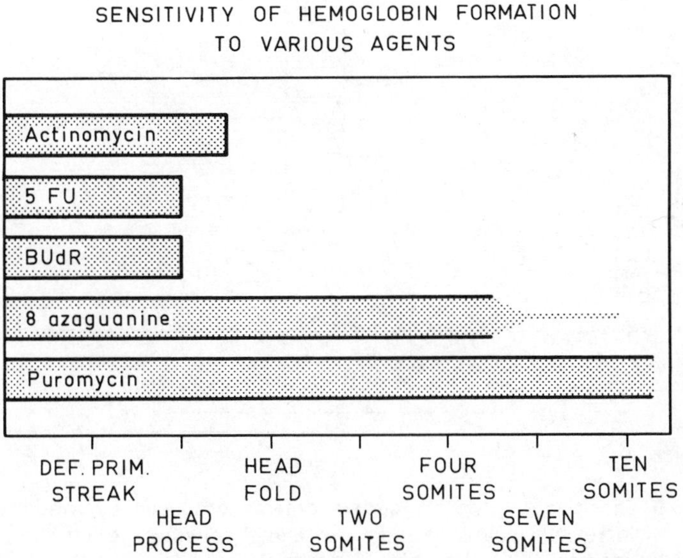

Fig. 5.8 Inhibition of hemoglobin formation in explanted chick blastoderms by different metabolic inhibitors. The shaded bars indicate the period of sensitivity to different compounds. (After F. Wilt. Amer. Zool. **6:** 67 [1966].)

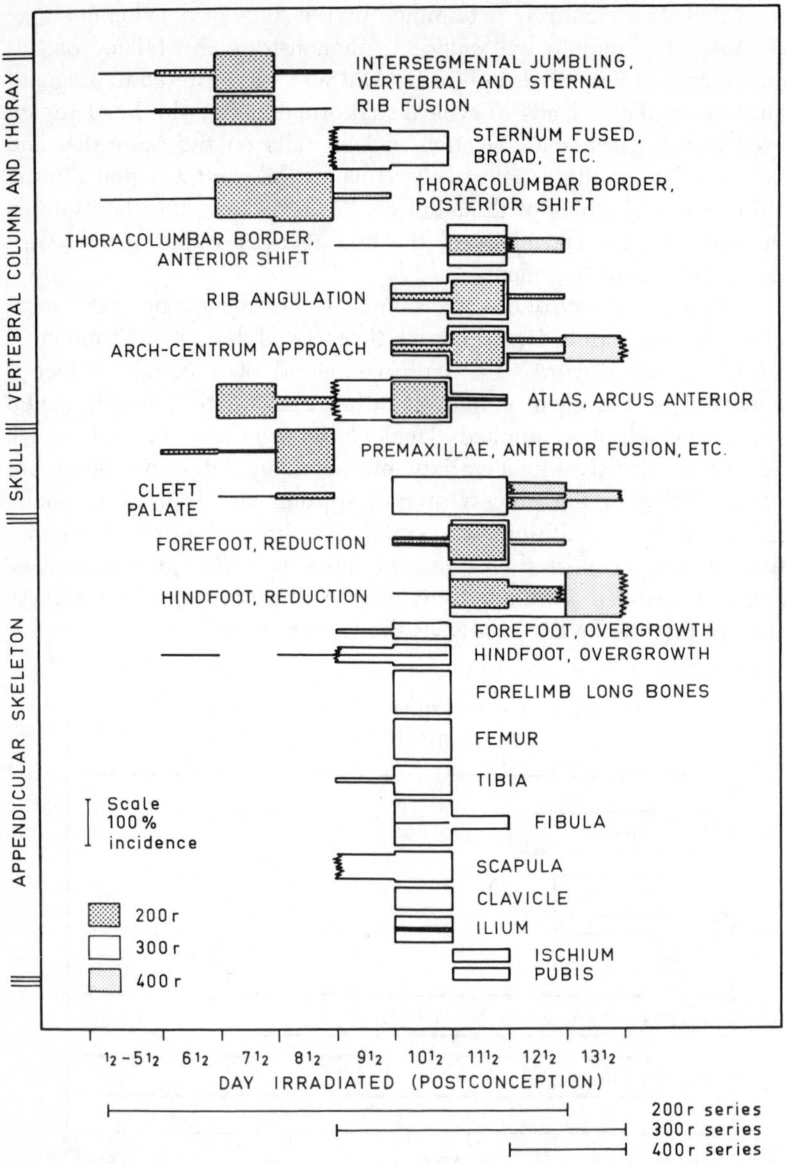

Fig. 5.9 Sensitive periods for certain defects produced by irradiation in mice. An increased incidence and an extended sensitive period can be obtained by increasing the intensity of irradiation (*see* cleft palate, hind foot reduction). (After L. B. Russell and W. L. Russell. J. Cell. Comp. Physiol.: **43** Suppl. 103 [1954].)

Dose-Dependence of the Sensitive Periods

In the examples of general and specific sensitive periods presented above, the dose used was constant and the treatment was similar at different stages. It should, therefore, be asked whether a change in the dose of the teratogen or the intensity of the treatment would have yielded different critical periods or whether these would have proved completely independent of the dose. If the latter were true, we would expect that whenever a teratogen reaches the threshold level for manifest effects, the prevailing epigenetic crisis would be influenced and an invariable effect be obtained, while developmental processes at less critical stages would continue unscathed. It is hardly surprising, however, that this is not the case and experimental data indicate that an increase of the dose or intensified treatment leads to extension of the sensitive period for a given defect. In addition, new defects will be produced during the same period and the incidence of defects obtained with lower doses will increase (Fig. 5.9).

The results concerning the induction of cleft palate shown in Fig. 5.9 deserve special attention. Radiation of 200 *r* produces this defect during day 7 or 8, but if the dose is increased to 300 *r*, a second critical period is revealed around day 10 or 12. Moreover, when the treatment is further intensified, the latter sensitive period is extended by a day or two. Similar results have been obtained with different chemical teratogens and the bearing of such observations on the whole problem of epigenetic crises will be discussed at the end of this chapter.

Interstrain Differences in the Sensitive Period

As already stressed, the response of an embryonic organism to an exogenous teratogen is largely dependent on the developmental stage of the target organism. Consequently, if genetically determined differences in the timing of normal embryogenesis exist, one would expect dissimilarities in the sensitive periods. Examples of such interstrain differences in the developmental age as compared with the chronological age are, in fact, known (see page 83, and Dagg, 1966) and it is tempting to correlate these to the known differences in the sensitive periods. As an experimental demonstration of such strain-dependent differences, the work of Dagg (1960) may be quoted (Table 5.3). When 5-fluorouracil was administered to embryos of two inbred strains of mice (129/c and BALB/c), certain differences in their sensitive periods were noted. The difference was best seen in the time of maximal tendency of the drug to induce limb malformations and the author concluded that the sensitivity peak occurred one day later in strain BALB. It is of further interest to notice that these experiments, too, like those illustrated in Fig. 5.9, indicate a bimodal sensitive period for the induction of cleft palate.

TABLE **5.3**
Sensitivity (x) and Period of Maximal Sensitivity (xx) of Certain Malformations Produced in Two Strains of Mice by Treatment with 5-Fluorouracil at Different Stages of Development.[a]

Defect	Age of Embryo at Treatment						Strain
	9	*10*	*11*	*12*	*13*	*13.5*	
Cleft		xx	x	x	xx		129
Palate		x	x	x	x		BALB
Enlarged		xx	x				129
Hind Foot		x	xx				BALB
Reduced	x	x	xx	x		x	129
Hind Foot		x	x	xx			BALB
Reduced			xx	x	x		129
Fore Foot			x	x			BALB

[a] From C. P. Dagg: Am. J. Anat. **106:** 89–96, 1960.

FACTORS DETERMINING SENSITIVE PERIODS

The many examples presented above allow us to conclude that there are definite differences in the susceptibility of embryonic tissues and of whole embryos at different stages of development. Moreover, they show that different organs may be affected at different times, which are determined by the dose of the teratogen and the genetic constitution of the animal. In the following paragraphs, an attempt is made to analyze these different factors and to speculate about the different mechanisms possibly determining these periods. In addition, it is suggested the reader refer to a similar discussion in Chapter 9, where the specific problem of virus susceptibility and its control will be discussed.

Epigenetic Crises

The expression "epigenetic crises," coined by Waddington, has already been referred to in suggesting that certain events of normal development are more vulnerable than others and that a developing organism passes through a succession of critical periods liable to be affected by environmental factors. As an example, we may take gastrulation. During this stage large masses of tissue migrate to new positions, a process easily prevented by various treatments (pH, temperature, mechanical manipulation, chelating agents, and so forth). Not only do such treatments prevent the occurrence of a particular developmental event but, since gastrulation is of major importance for the future determination of the central nervous system, the consequences are profound, as may be seen from Fig. 4.1.

Yet we can only surmise how many such morphogenetic movements followed by inductive tissue interactions lead to the onset of new synthetic activity vulnerable to environmental factors. One example will be dealt with here in some detail to show the complexity of the situation even in a simplified model *in vitro.* As already described on page 31 and in Fig. 2.14, the inductive interactions leading to the determination of kidney tubules have become a useful tool for certain teratologic analyses. The inductive stimulus triggering this morphogenetic event can be exactly timed *in vitro* and hence the subsequent course of events in different series of experiments can be temporally correlated. Figure 5.10 shows a series of experiments in which the effect of Actinomycin D on this system was tested. This treatment, often referred to as "chemical enucleation," prevented the formation of kidney tubules, if applied during a sensitive period lasting some 24 hours subsequent to induction. Yet in untreated explants the process of tubule formation can only be detected about 12 hours later. Consequently, we may state that in this case a "silent" period intervenes between the sensitive period and the overt signs of the event that is to become affected.

The same series of experiments demonstrates another point of interest. The sensitive period in the very same cells is different for different developmental events apparently induced by the same primary trigger. In this particular case, chemodifferentiation was followed by determining the patterns of lactate dehydrogenase in the responding tissue. During normal differentiation a shift of these isozymes from the cathodal to the anodal type reflects chemodifferentiation at the protein synthesis level. This shift was prevented by treatment with actinomycin, but the sensitive period was definitely longer for this effect than for prevention of morphogenesis. Different explanations can be suggested and one can speculate on sequential activation of the genome (inhibited at different stages by actinomycin). Subsequent to the sensitive period, the message has been delivered to the synthesizing machinery and can no longer be prevented by treatment affecting the flow of information from the nucleus to the cytoplasm.

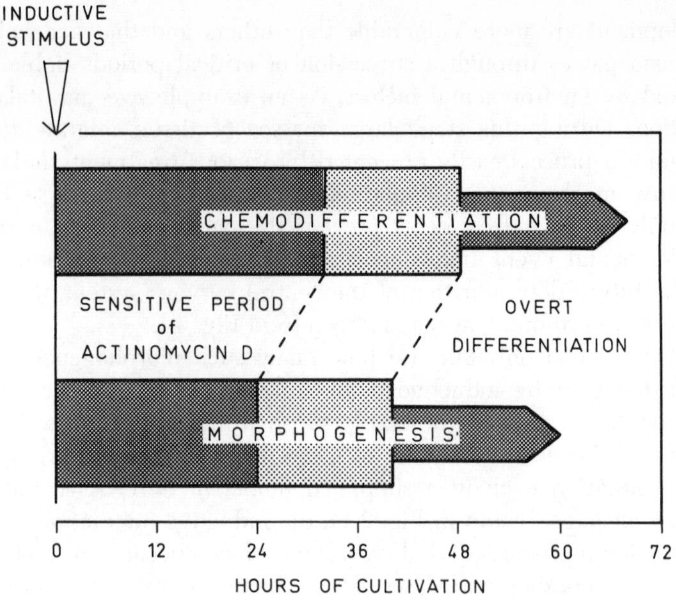

Fig. 5.10 The "sensitive periods" of Actinomycin D during differentiation of kidney tubules *in vitro*. Chemodifferentiation was followed as changes in the lactic dehydrogenase pattern and the morphogenesis as differentiation of tubule structures. (After O. Koskimies. Exp. Cell. Res. **46:** 541 [1967].)

When the concept of epigenetic crises is pursued further, it appears only natural that a certain mature stage will have to pass through several such tight corners in the course of its development. This, in turn, would explain the existence of two or more sensitive periods for the same defect, as previously shown in the experiments by Russell and Dagg on cleft palate. A detailed study of such sensitive periods and their temporal correlation to major developmental events will certainly show us a number of epigenetic crises, but we have to remember that the sensitive period does not necessarily coincide with an overt morphogenetic event. The teratogen may exert its effect during a "silent" period, long before differentiation is detectable, as shown in Fig. 5.10.

Restoration Capacity

As mentioned in the beginning of this chapter, embryogenesis seems to be well canalized and both genetic and environmental factors buffer normal differentiation against harmful environmental factors.

The following example may illustrate one such restoration capacity of embryonic cells. If a tissue anlage consisting of more than one cell type is completely disaggregated and the cells thoroughly mixed, they will spontaneously reaggregate *in vitro* and continue growth as a dense colony. Inside this cell mass, the different randomly moving cell types will recognize their own kind and adhere to them more closely than to cells with a different developmental history, so that subsequently these aggregates of like cells sort themselves out and form homogeneous colonies. Moreover, these colonies of like cells become segregated according to their normal sequence and in a disaggregated gastrula we soon find a central entoderm surrounded by a mesodermal mantle, which in turn is covered by epidermis (Townes and Holtfreter, 1955). The drastic treatment leading to complete disruption of the normal architecture thus has not demolished the developmental capacities of the cells and will be followed by almost normal development. Similarly, large portions of an undetermined embryo can be mechanically removed without affecting development; the area will be replaced by new undertermined cells, which subsequently become exposed to the control system of normal development and complete embryos eventually develop. On the other hand, if cells already determined are removed, the restoration capacity, that is, the flexibility of the organism, is already weaker and a similar operation during a later stage will lead to a corresponding defect. If, furthermore, the part to be removed or destroyed normally exerts some control activity (being itself an inductor) the defect will not be restricted to the destroyed area, but will manifest itself as a spread of the effect due to incomplete determination of the neighboring cells.

The role of the regeneration capacity and the progressive increase of determination to the teratogenic action of environmental factors is schematically illustrated in Fig. 5.11. During the early stages, a vulnerable treatment usually results in the destruction of relatively large parts of the embryo (cells in the same stage of differentiation and with similar thresholds). This original lesion can thereafter develop in three different ways, as shown in Fig. 5.11. It may resolve completely, or it may spread and lead to death of the embryo or to a manifest defect. This may be the situation in blastogeny and all these end results have been produced experimentally. During subsequent differentiation, the number of undetermined cells decreases and the regeneration capacity is gradually lost, but because of the specialization of the cells, the number of these cells at a critical stage is limited and the primary lesion correspondingly restricted. Consequently, the treatment seldom leads to complete destruction of the embryo and the outcome is either a defect proportional to the original destruction or a more generalized effect due to spread of the lesion

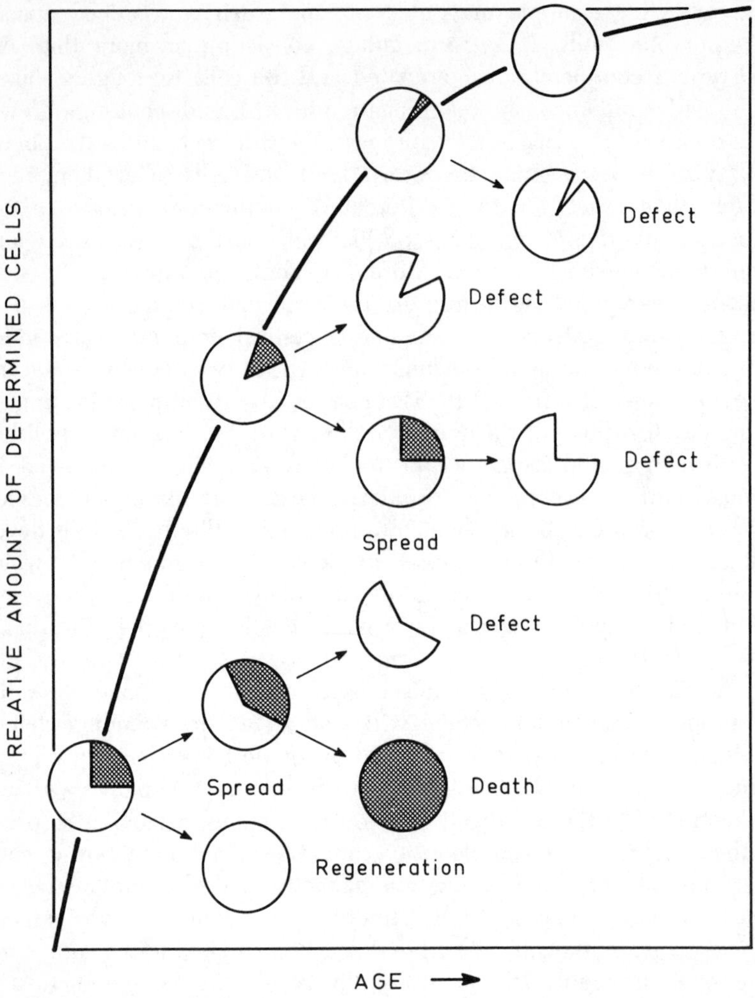

Fig. 5.11 A hypothetic scheme of the possible consequences of teratogenic treatment as a function of age and the relative amount of differentiated tissue.

mentioned above. This stage, perhaps, may be correlated to the period of major organogenesis and is followed by a stage of still more restricted defects, but also by restricted regenerative capacities.

Barriers

Before producing an effect on the developing organism, a teratogen, unless acting through the maternal organism, has to enter the target tissue. In most experimental and clinical conditions this takes place through the placenta and subsequently through fetal barriers (endothelium of the vessels) and cellular membranes. Each of these may, in fact, alter its permeability during development and thus contribute to the development of relative resistance, the end of the sensitive period. Very little is known about these barriers, but the placenta might, in fact, be a somewhat overrated barrier which is now thought to exclude from the embryo little except very large or highly charged molecules. In addition, very many of the experiments described above have obviated the participation of such barriers, either by using treatment independent of them (irradiation) or by observing the teratogen in the fetal organism. Moreover, the *in vitro* studies illustrated in Fig. 5.10 were accompanied by direct measurement of the uptake of the drug by tissue during and after the sensitive period without detecting any differences in its penetration. Still, true barriers and the development of either mechanical or chemical defense mechanisms have to be considered in many instances until definitely excluded. Later, in Chapter 9, we will discuss the penetration of virus particles into the fetal organism.

Proliferative and Metabolic State

Finally, changes in the susceptibility at the cellular level have to be mentioned. Many *in vitro* studies not directly related to teratogenesis have shown that the proliferative and metabolic state of a cell may have a bearing on its susceptibility to environmental factors like radiation, drugs, and viruses. A general impression in teratology seems to be that the teratogen is more likely to affect embryonic cells of high metabolic and proliferative activity, whereas cells in a resting phase may remain intact. Whether such factors really represent an epigenetic crisis at the cellular level remains to be settled, but some examples may be quoted here. Regarding the proliferative stage, both drugs and certain viruses (Margolis *et al.*, 1967) seem preferentially to attack mitotic cells, and the experiments by Hemsworth and Jackson (1965) may serve here as an example. They showed that Busulphan, an alkylating agent, when given to pregnant rats, causes congenital sterility in their offspring. Both male and female fetuses are sensitive to the sterilizing activity of this drug, but at different stages of development corresponding to the time when their primitive germ cells divide (Fig. 5.12). The sterilizing effect of Busulphan in the female embryo ends at the time when the oocytes stop dividing in the first period. Spermatogonia, on the other hand, in accordance with their continuous proliferation, are susceptible throughout the latter part of gestation.

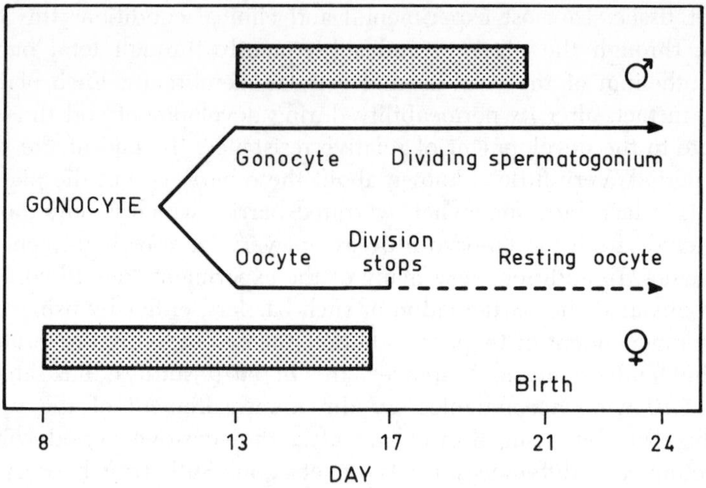

Fig. 5.12 The time of the sterilizing effect of Busulphan (shaded area) is related to the proliferation of the germ cells in both sexes. (After B. N. Hemsworth and H. Jackson. *In* Embryopathic Activity of Drugs, J. M. Robson, F. M. Sullivan and R. L. Smith [eds.], Churchill, London [1965] p. 116.)

Irradiation, on the contrary, does not necessarily destroy proliferating cells in the first place. In the experiments illustrated in Fig. 7.5, the most sensitive cells seemed to be the primitive neuroblasts and spongioblasts. (The sequential effect of irradiation on the different parts of the central nervous system was mainly attributable to the presence of these sensitive cells at the time of treatment.) Yet these cells have a relatively low mitotic activity, whereas their rapidly dividing precursors, the neuroectodermal cells, were more resistant to the treatment.

"High metabolic activity" is naturally a term easy to apply but rather difficult to define or expose to experimental testing of its greater vulnerability than "low metabolic activity." In the case of the kidney induction shown in Fig. 8.7, both nucleic acid and protein synthesis were at a low level during the sensitive period and only increased subsequent to it. Apparently, the term should be applied to certain specific metabolic pathways inhibited by the teratogen and necessary for subsequent differentiation. The questions bring us back to the problem of thresholds already discussed in connection with the synergistic effects of genetic and environmental factors (page 56). We may close this discussion by a brief repetition of the interesting hypothesis of Rutter *et al.* (1967) and its application to some teratologic studies already presented. According to this hypothesis, differentiation consists of two (or more) steps, the first of which involves reorganization of the genetic expression and the second, regulation of the activated genome. In his model system (pancreatic rudiment), the first stage, protodifferentiation, is characterized by the onset of synthesis of specific proteins, although at a relatively low level. During active differentiation, a secondary regulative system raises these synthetic activities to a level at which the differentiation of the pancreatic cells is manifest. The first stage is actinomycin-sensitive (compare to Fig. 5.10) but a less sensitive period soon supervenes, indicating that additional messenger is no longer required for manifest differentiation. In addition, the first step toward differentiation, protodifferentiation, is accompanied by increased mitotic activity, which ceases during the second stage. We may now speculate on this model as an example of the different sensitive periods, the role of the time of treatment, and the specificity of teratogens applied at different stages in the course of this development. In doing so, we postulate that all the inhibitory treatments employed in our theoretic experiment lead to an identical end result—inhibition of pancreatic differentiation.

During "protodifferentiation" at least two basically different treatments would lead to the same end result, treatment with mitotic inhibitors and/or application of compounds preventing the synthesis of messenger-type RNA. Sensitivity to both these treatments ends at the time when mitotic activity normally ceases and the messenger required for subsequent differentiation has been synthesized. Here the sensitive periods of the two treatments may be identical, or at least overlap, without meaning that the mechanism of their effect is similar.

During "actual differentiation" the above mentioned teratogens are no longer harmful, but some new factors may become involved. The activation of protein synthesis may be prevented by specific metabolic inhibitors, the actively synthesizing cells may be comparable to the neuroblasts mentioned above and be highly susceptible to irradiation, and so on. Consequently, new sensitive periods will be created for the new teratogens, all leading to the same end result, as suggested above.

Finally, we may tackle the problem of the dose-dependence of the sensitive period already shown in Fig. 5.9. If we suppose that the manifest process of differentiation is a reflection of new synthetic activities gradually increasing to the threshold levels necessary for permanent specialization of the cells, we can speculate on the significance of increasing doses of a teratogen (Fig. 5.13). A low dose may only be able to inhibit a low level of a certain activity and once this has been reached, such a dose would no longer be sufficient to prevent the subsequent increase to the threshold level (*a* in Fig. 5.13). The same might be true with a somewhat larger dose, still sufficient to inhibit increased activity during later stages (prolonged

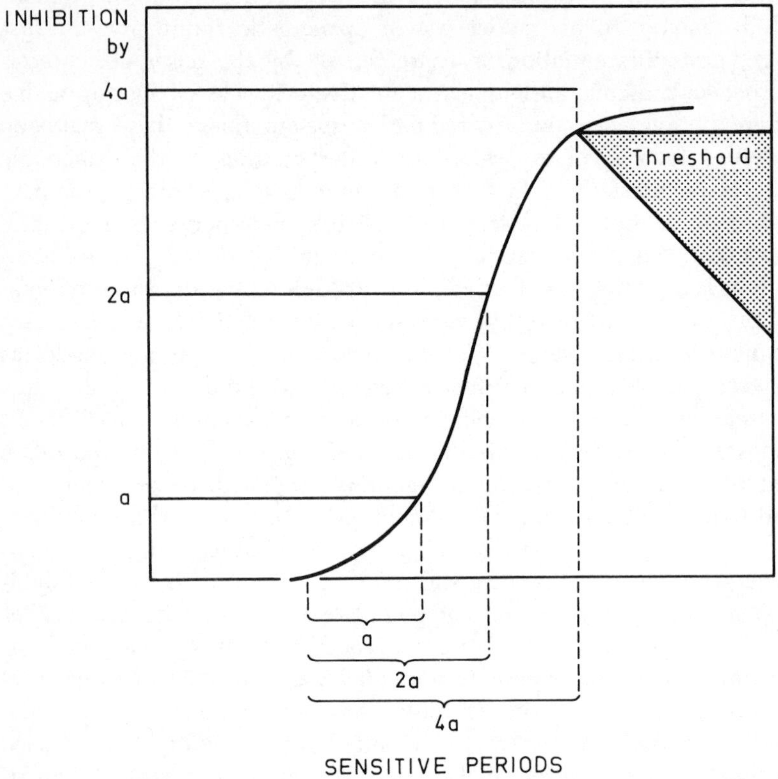

Fig. 5.13 A speculative scheme of the effect of increasing doses of a teratogen on the extension of the sensitive period.

sensitive period; *2a* in Fig. 5.13), whereas a dose adequate to suppress the activity up to the threshold value would be active until the threshold of ultimate differentiation had been reached (*4a* in Fig. 5.13). We may stress that the activity curve need not necessarily be confined to some synthetic activity of the target cells, but may equally well refer to other features essential for differentiation, such as mitotic activity, cellular motility, adhesiveness, and so on. Direct evidence for this hypothesis is still lacking, however.

FETOPATHIES

As major organogenesis is completed relatively early in intrauterine development, the latter part of fetal life is a period of growth and maturation. Consequently, no malformations in the sense of developmental arrests would be expected except in those organs still undergoing differentiation (central nervous system). In a wider sense, however, congenital defects can still be induced, provided that the immature organism differs from the adult in its susceptibility to a certain teratogen. In addition, some mechanical factors may come into play in the uterus. On the following pages, we briefly discuss some conditions harmful to developing fetus during late pregnancy.

The reader should also consult Chapter 6, where the endocrine problems and hormonal homeostasis during late pregnancy will be discussed in greater detail. In addition to some passing comments, three examples of "fetopathic" factors have been singled out for presentation: one representing a late fetal infection; one, a harmful drug effect; and the third, abnormal mechanical conditions in the uterus possibly leading to maldevelopment in the wider sense.

Toxoplasmosis

In addition to some virus infections causing fetal diseases during late pregnancy (Chapter 9), toxoplasmosis may affect the embryo at any stage of development. This often subclinical or even latent maternal disease is caused by the protozoan *Toxoplasma gondii,* an organism about $5 \times 3\,\mu$ in size. Probably by way of the bloodstream, the parasite gains access to the placenta, where it causes necrobiotic changes and may enter the fetus. In man, a recent infection during pregnancy is estimated to result in infection of the fetus in 40 to 50 percent of cases (Frenkel, 1967) and transplacental transmission has been demonstrated experimentally in a variety of laboratory animals. Depending on the stage of pregnancy at which fetal infection takes place, different consequences may be expected. Early contamination leads to fetal death (experimentally in mice and probably in man), while infection during the organogenetic stage may lead to malformations or persistent infection and, finally, fetal contamination during late pregnancy causes "congenital toxoplasmosis," a condition best known from human pathology. The typical triad for this disease consist of an inflammatory change in the eye (chorioretinitis), hydrocephalus, and intracerebral calcifications detectable with x-rays. The latter two symptoms seem to justify us in placing the condition under the title "congenital defects" since the hydrocephalus, at least, closely resembles a true developmental disturbance. The mechanism appears to be the destructive changes brought about by the intracellular multiplication of the parasite, with destruction of brain tissue, the resultant granulation, and fibrosis blocking the outlets of the ventricular system. (Correspondingly, calcium is deposited in the small necrotic foci.) Intrauterine infection may lead to the death of the fetus, but more frequently to the birth of a congenitally diseased child. At present, we do not know the magnitude of the risk of defects following a maternal toxoplasma infection, but several well-analyzed cases have convincingly showed that a causal relationship does exist (for example, Miller *et al.*, 1967).

Tetracycline

In Chapter 2 we stated briefly that the antibiotic tetracycline is selectively incorporated in calcifying tissues and may interfere with osteogenesis. In addition, we know that the drug will readily pass the placenta of various experimental animals and pregnant women and is soon detectable in the fetal skeleton. Its inhibitory effect on mineralization has been shown in certain lower vertebrates and in mammalian bones cultured *in vitro* (Bevelander, 1964; Saxén, 1966), but the situation *in utero* seems to be more complex.

As stated, tetracycline can easily traverse the human placenta and in some cases its bright fluorescence has been seen in the skeletons of stillborn children of mothers who had received intense tertracycline treatment during late pregnancy. In surviving children, a consequence is the brownish and permanent discoloration of the teeth and enamel hypoplasia has even been suggested. Moreover, tetracycline treatment of premature children has been shown to cause significant, although transient and reversible, retardation of mineralization (Cohlan *et al.*, 1963). Hence, ample indirect evidence is available to show that tetracycline is a late teratogen operating during the second half of pregnancy. The drug passes the placenta, is detectable in the bones, and causes definite inhibition of mineralization in the lower vertebrates, in mammalian bones *in vitro,* and in premature children. Direct evidence of its harmful effect on human fetuses during late pregnancy is still lacking.

Among other drugs and active compounds exerting harmful transplacental effects during late pregnancy, the following may be listed: progesterones causing masculinization of female fetuses; antithyroid drugs and radioactive iodine causing congenital goiter; vitamin K causing bilirubinemia; and salicylates causing hemorrhages.

Mechanical Factors

During late fetal life, the developing organism may be exposed to abnormal spatial conditions or to mechanical injuries due to abnormalities of the uterus or the membranes. Browne (1967) has recently listed the four major categories of such conditions with four hypotheses: malposition, due to lack of space or an abnormal amount of space allowing extended movements; increased spatial pressure due to undersized mother, oligohydramnion, and so on; increased hydrostatic pressure, due to hydramnion, interference with circulation, and so on; and finally perforation of the membranes by the fetal extremities. The above list, like most previous ones, is mainly concerned with a group of limb defects, talipes, "in which all normal elements are present and which are deformed in a way that *could* have been produced by moulding before birth." These various defects, common in human populations, comprise abnormalities in the position of the legs followed by profound remodelling of the bony elements. To judge by their appearance, a mechanistic explanation of their etiology sounds plausible. As indirect evidence, we may add that structurally similar deformations can be produced in newborn animals by mechanical means. General agreement has not yet been reached concerning the part played by such mechanical factors in maldevelopment during late pregnancy and, in addition, one should ask whether, here again, we are dealing with malformations or with deformations. In practice, however, most statistics seem to include these limb defects under the general title "congenital malformations."

We may close the section on fetopathies by stating that, after passing the period of active organogenesis, the embryo may be protected against most teratogens, but is still vulnerable to a variety of exogenous factors, some of which were listed above. Consequently, the sensitive period in a wider sense should be extended to cover the whole intrauterine period—both in theory and in practice.

SUGGESTED READINGS

Review Papers

Braden, A. W. H. Are nongenetic defects of the gametes important in the etiology of prenatal mortality. Fertility Sterility **10**: 285–298 (1959).

Dagg, C. P. Teratogenesis. *In* Biology of the Laboratory Mouse (E. L. Green, ed.), pp. 309–328. McGraw-Hill, New York (1966).

Degenhardt, K. H. Kritische Phasen der Musterbildung in der Frühentwicklung des Menschen. Verhandl. Ges. Deut. Naturforsch. Ärzte **103**: 186–199 (1964).

Hamilton, H. L. Sensitive periods during development. Ann. N.Y. Acad. Sci. **55**: 177–188 (1952).

Thalhammer, O. Pränatale Erkrankungen des Menschen. Georg Thieme Verlag, Stuttgart (1967).

Töndury, G. Die kritischen Phasen in der Embryonalentwicklung und ihre

Störung durch chemische Faktoren und Viren. Vjschr. Naturforsch. Ges. Zürich. **101**: 93–138 (1956).

Waddington, C. H. Tendency towards regularity of development and their genetical control. *In* International Workshop Teratol. pp. 66–75, Copenhagen (1966).

Witschi, E. Overripeness of the egg as a cause of twinning and teratogenesis. A review, Cancer Res. **12**: 763–786 (1952).

Special Articles

Adams, C. E., M. F. Hay, and C. Lutwak-Mann. The action of various agents upon the rabbit embryo. J. Embryol. Exp. Morphol. **9**: 468–491 (1961).

Bevelander, G. The effect of tetracycline on mineralization and growth. *In* Advances in Oral Biology, (P. H. Staple, ed.), vol. 1, pp. 205–223. Academic Press, New York (1964).

Browne, D. A Mechanistic interpretation of certain malformations. *In* Advances in Teratology, (D. H. M. Woolam, ed.), vol. 2, pp. 11–36. Academic Press, London (1967).

Butcher, R. L. and N. W. Fugo. Overripeness and the mammalian ova. II. Delayed ovulation and chromosome anomalies. Fertility Sterility **18**: 297–302 (1967).

Cohlan, S. Q., G. Bevelander, ja T. Tiamsic. Growth inhibition of prematures receiving tetracycline. Am. J. Diseases Children **105**: 453–461 (1963).

Dagg, C. P. Sensitive stages for the production of development abnormalities in mice with 5-fluorouracil. Am. J. Anat. **106**: 89–96 (1960).

Ferm, V. H. Developmental malformations as manifestations of reproductive failure. *In* Comparative Aspects of Reproductive Failure, (K. Benirschke, ed.), pp. 246–255. Springer-Verlag, New York (1967).

Frenkel, J. K. Toxoplasmosis. *In* Comparative Aspects of Reproductive Failure, (K. Benirschke, ed.), pp. 296–321. Springer-Verlag, New York (1967).

Gottschewski, G. H. M. and Zimmermann, W. Nachweis von phänokritischen Phasen verschiedener Organanlagen beim Hauskaninchen, *Oryctolagus cuniculus*. Verhandl. Deut. Zool. Ges. München, pp. 144–176 (1963).

Hertig, A. T. The overall problem in man. *In* Comparative Aspects of Reproductive Failure, (K. Benirschke, ed.), pp. 11–41. Springer-Verlag, New York (1967).

Lutwak-Mann, D. Observations on progeny of thalidomide-treated male rabbits. Brit. Med. J. **I**: 1090-1091 (1964).

Lutwak-Mann, C., K. Schmid, and H. Keberle. Thalidomide in rabbit semen. Nature **214**: 1018–1020 (1967).

Margolis, G., L. Kilham, and J. Davenport. A model for virus induced reproductive failure: Theory observations and speculations. *In* Comparative Aspects of Reproductive Failure, (K. Benirschke, ed.), pp. 350–360. Springer-Verlag, New York (1967).

Miller, M. J., E. Seaman, and J. S. Remington. The clinical spectrum of congenital toxoplasmosis: Problems in recognition. J. Pediat. **70**: 714–723 (1967).

Rugh, R. and M. Wohlfromm. Can the mammalian embryo be killed by X-irradiation. J. Exp. Zool. **151**: 227–240 (1962).

Rutter, W. J., W. D. Ball, W. R. Bradshaw, W. R. Clark, and T. G. Sanders. Levels of regulation in cytodifferentiation. *In* Morphological and Biochemical Aspects of Cytodifferentiation, (E. Hagen, W. Wechsler and P. Zilliken, eds.) Experimental Biology and Medicine, vol. 1, pp. 110–124. S. Karger, Basel (1967).

Saxén, L. Effects of tetracycline on osteogenesis *in vitro*. J. Exp. Zool. **162**: 269–294 (1966).

Townes, P. I. and J. Holtfreter. Directed movements and selective adhesion of embryonic amphibian cells. J. Exp. Zool. **128**: 53–120 (1955).

6

Endocrines and Maldevelopment

The endocrine system brings the different parts of the animal organism into mutual functional contact and regulates many of the vital functions of animals. Reproduction, particularly, is controlled in all higher animals by hormones. It is well known that the development and maturation of gametes, ovulation and in placental animals transport of ova, and preparedness of the endometrium for pregnancy are all influenced by hormones. Furthermore, mating behavior and the maintenance of pregnancy are at least partially regulated by endocrine mechanisms and there is evidence that termination of pregnancy and onset of parturition are accompanied by hormonal changes.

The role of hormones in development does not end with these indirect regulatory mechanisms, but hormones also participate in the differentiation and maturation of fetal tissues. The mammalian embryo is exposed to a complex and changing hormonal environment during its intrauterine development. Hormones are supplied from the mother, from the fetus itself, and from the placenta. A disorder in one part of this triple system may extend its effect to the other parts and so disturb the hormonal balance needed for normal development.

The purpose of this chapter is not to give a full survey of fetal endocrinology. For such a survey, the reader is referred to the *Role of Hormones in Development* by O. Hamilton, in this series of publications on developmental biology. Clinical aspects of fetal endocrinology are dealt with in several recent surveys, for example, of Liu (1966).

ONTOGENY OF THE ENDOCRINE ORGANS AND HORMONAL FUNCTION

The endocrines begin their organogenesis independently of each other. Their early differentiation is regulated by nonhormonal mechanisms such as morphogenetic movements and tissue interactions, as has been experimentally shown in the case of the thyroid gland (Hilfer, 1962). The later development of the endocrine glands and the onset of specific activity in several glands are dependent on the synchronous interaction of the endocrine system.

Experiments performed during the early decades of the century with amphibian tadpoles showed that removal of the hypophyseal primordium prevented the development of the peripheral endocrine glands and metamorphosis, which is the most conspicuous hormone-dependent course of developmental events (see Allen, 1938). Later, the ingenious intrauterine experimental procedures of Jost and his collaborators (see Jost, 1953, 1961, 1966) and Wells (see Wells, 1965) have elucidated the role of hypothalamic-hypophyseal function in the control of the functional development of the peripheral endocrine glands. The results of these experiments may be summarized as follows: All peripheral endocrines start their function during the latter half of prenatal development. Their functional maturation, with some exceptions, is dependent on the hypothalamic-hypophyseal tropic action. The feedback mechanism which supresses the hypothalamic action is exerted from the peripheral glands and is also functional before birth.

It seems that under normal conditions the fetal hormones themselves are the most important factors regulating the development of the fetal endocrines, the maternal hormones playing only a minor role. Certain disturbances in maternal endocrine status, however, are reflected in fetal hormone balance, as will be seen later.

PLACENTA AND FETO-PLACENTAL UNIT

The placenta is a transient gestational organ of mammals, necessary for the survival and development of the mammalian fetus. It is logical to suppose that failures in its function will lead to fetal death or developmental defects, but very little direct evidence that its malfunction causes congenital defects has been gathered. Some examples of the indirect role of the placenta in the genesis of exogenously caused defects are discussed in Chapter 8, page 186.

One of the significant functions of the placenta is to produce hormones during pregnancy. Thus it is known that in some species, including man, the placenta produces the hormones responsible for the maintenance of pregnancy, luteotropic hormone, progesterone, estrogens, and a hormone, placental lactogen, which presumably prepares the mother for lactation (see Josimovich, 1967; Ryan and Aimsworth, 1967).

One aspect of placental function related to prenatal endocrinology will now be discussed, showing the intimate interaction of the fetal organism and placenta in prenatal hormone production. Recent studies have shown that the placenta produces steroid hormones—progesterone and estrogens—in large quantities, but is not able to synthesize the steroid nucleus from simple precursors. The main precursor for progesterone synthesis, cholesterol, is supplied by the mother, but estrogen synthesis is the result of fetal and placental coaction, called the feto-placental unit (see Klevit, 1966; Mitchell, 1967). Steroid hormone biosynthesis occurs in the glands, where the precursor steroid ring is changed through consecutive chemical reactions to active hormones with the aid of enzymes. In the case of prenatal estrogen synthesis, part of the enzymatic function takes place in the fetal adrenal cortex, which begins its specific function at the fifteenth week of gestation. The fetal adrenal cortex, however, does not produce all enzymes in sufficient quantities, and the incomplete hormones, mainly dehydroepiandrosterone (DHA), are transferred to the placenta for completion of their synthesis. It is even possible that the steroid metabolite returns to the fetus for the final step of estriol synthesis (see Fig. 6.1). The key enzymes in the placenta are 3β-hydroxysteroid dehydrogenase, Δ^5isomerase, and aromatase. The fetal adrenal gland provides the initial steroid, DHA, and both the adrenal and the fetal liver have the high activity of 16α-hydroxylase needed for estriol synthesis, the latter being the main estrogenic hormone found in the urine of the pregnant mother. Neither placenta nor fetus can produce estriol in any appreciable quantities alone.

The biological significance of the high estriol production of the feto-placental unit is not yet understood, but it is an important diagnostic sign of fetal development. The rate of excretion of estriol in the maternal urine correlates well with fetal weight and viability during the latter half of pregnancy (Frandsen and Stakemann, 1963). Low values of estriol and particularly a sudden drop in its excretion indicate an emergency state of the fetus and, if permanent, are the most reliable biochemical markers of fetal death. As would be expected, low estriol values are found in cases of underdevelopment of the fetal adrenal gland. The commonest cause of adrenal underdevelopment is anencephaly. Thus, in the case of a living fetus, a low estriol value is a suggestive diagnostic criterion for the expected birth of an anencephalic infant (Frandsen and Stakemann, 1964).

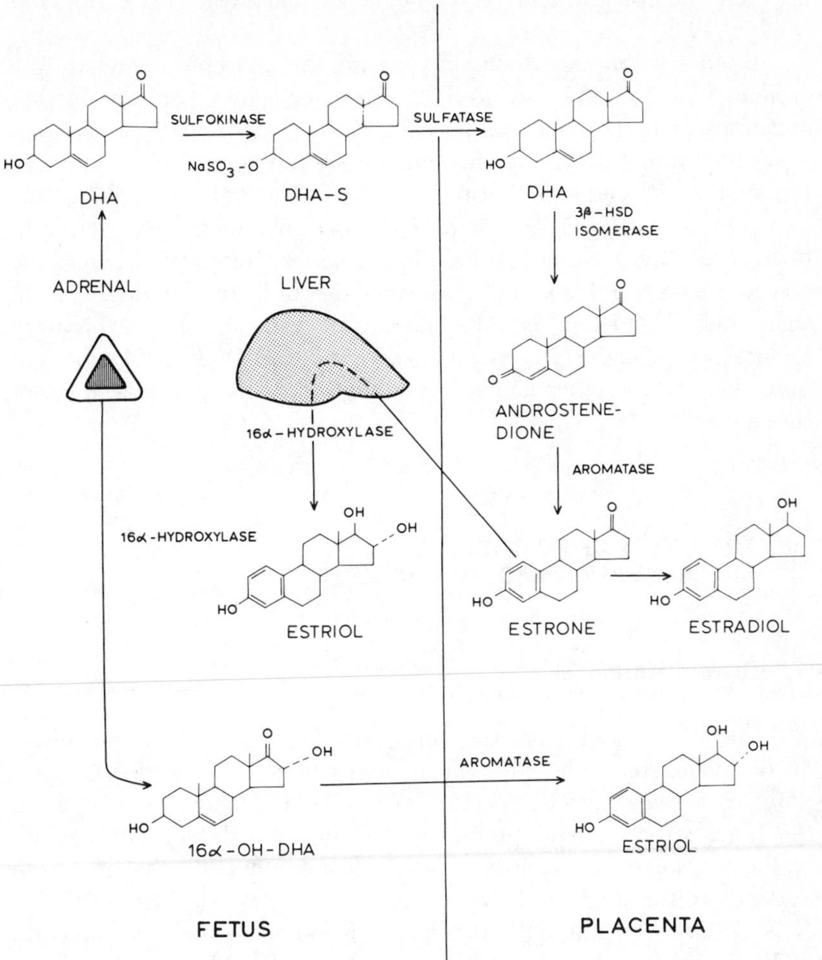

Fig. 6.1 Schematic presentation of estrogen synthesis in the feto-placental unit. DHA from the fetal adrenal is transferred to the placenta after sulfurylation DHS-S. After placental desulfurylation it is converted to estrogens through enzymatic steps and the hormone returns to the fetus, where hepatic or adrenal 16α-hydroxylase converts it to estriol. An alternative pathway, in which DHA is first hydroxylated to 16α-hydroepiandrosterone and then transferred to the placenta for completion of estriol synthesis, is also indicated. (Modified from N. Lu Pediat. Clin. N. Amer. **13:** 1047 [1966].

Sometimes, however, a living but not apparently malformed child is born even when the estriol excretion during pregnancy has been low. In a number of these cases postnatal development has turned out to be abnormal and some of these children suffer disturbances in their neurologic and psychologic development (Wallace and Michie, 1966).

Besides being an endocrine organ, the placenta regulates the exchange of hormones as well as other molecules between mother and fetus (see Hagerman and Villee, 1960). Differential permeability is of importance in the understanding of the role of the fetus and the mother in endocrine disturbances of development. Polypeptide and protein hormones do not pass the placenta under normal conditions. Free thyroxine and triiodothyronine are transported across the placenta in several species. However, the bulk of the thyroxine in the maternal and fetal circulations is protein-bound and in this form it does not pass the placenta (French and van Wyk, 1964). Steroid hormones, on the other hand, seem able to pass the placenta in either direction.

THE EFFECT OF HORMONES ON DIFFERENTIATION AND DEVELOPMENT

Thyroid Hormones

The oldest and most thoroughly studied example of the effect of hormones on differentiation is amphibian metamorphosis. Amphibian metamorphosis consists of a wide array of changes in the body structure and physiology of the animal. Most, if not all, of these changes are caused by thyroid hormones secreted in large amounts in the thyroid of the metamorphosing animal (Etkin, 1955). It is remarkable that both regressive phenomena, such as reduction of the tadpole tail, and progressive phenomena, like the growth of limbs and many metabolic changes should be under the control of the same hormone, thyroxine (Weber, 1965). Even in the same organ thyroxine induces simultaneous regression and proliferation, as has been shown in the brain cells of tadpoles by Weiss and Rosetti (1951). It seems that in the metamorphosing cells of the frog, thyroxine triggers a preprogrammed mechanism, which is present dormantly in the premetamorphic cells. The responses of the different cells may be seemingly antithetic, but they are always consistent with the transformation of the tadpole to the adult frog.

In higher animals less is known about the developmental effects of the thyroid and many other hormones. Probably the best studied example of hormonally regulated differentiation is the development of the internal and external sex organs in higher animals. Some basic knowledge of the normal course of events greatly facilitates understanding of the pathology of sex differentiation.

Hormonal Regulation of Sex Differentiation

The genetic sex of animals, including man, is determined at fertilization, when the apportionment of sex chromosomes in the zygote dictates whether the new individual will be male or female. After this, however, many epigenetic control factors, mainly hormones, come into the picture and any major deviation among these may lead to abnormal development. In most cases the abnormality is expressed as development of sexual characteristics resembling those of the opposite sex.

In the lower animals complete sex reversal can be produced experimentally. With hormone treatment it is possible to change the sex of several amphibians and some fishes to the extent that the affected animals can breed normally according to their acquired, reversed sex (see Gallien, 1965). In higher animals, complete sex reversal is not possible, but even in mammals the development of the internal and external reproductive organs is under the control of hormonal factors.

Up to a certain stage of their development, the sex organs of an embryo are indifferent, that is, potentially able to develop into either sex. Diversification begins first in the gonads, then in the internal reproductive organs, and finally in the external sex organs. The embryonic organism has all the primordia needed for the complete development of either a male or a female.

Alfred Jost (1961, 1965), with elegant intrauterine surgical operations, has shown the dependence of sex organ differentiation on the fetal gonads. According to him and other investigators, it is the testicles that mainly determine the differentiation of the reproductive tract. Their action can be schematically divided into the following effects:

1. Degeneration of the Müllerian ducts
2. Stimulation of the Wolffian ducts to form the ductus deferens
3. Formation of the prostatic gland and seminal vesicles
4. Elongation of the genital tubercle and midline fusion of the genital folds and swellings to form the penis and scrotal sacs.

If the gonads are removed from the embryo at the neutral stage of development, the reproductive tract develops in the female direction, regardless of the genetic sex (Fig. 6.2). The Müllerian ducts persist, but the Wolffian ducts degenerate and the external genitalia

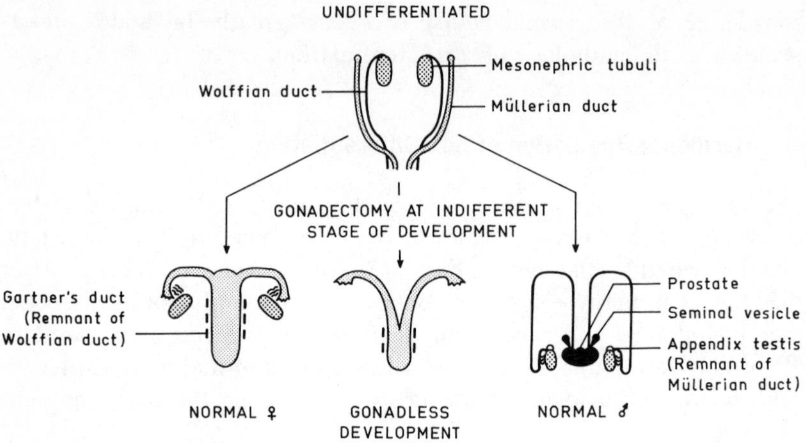

Fig. 6.2 Schematic representation of the development of the internal sex organs from the indifferent stage to male and female organs and the result of gonadless development (in middle).

also grow to resemble more or less those of the female in gonadless development.

Jost (see 1965) has presented evidence that two regulative principles are produced by the embryonic testis. The first "regulator" acts locally and induces degeneration of the Müllerian ducts. It is probably not a hormone, but is similar in nature to inducers which play a role in several morphogenetic interactions (see Chapter 4). The second principle, which is responsible for the maintenance and maturation of the Wolffian ducts and the development of the external genitalia and prostatic gland, consists of androgenic hormones similar to the androgenic hormones of the adult testis. This two-principle theory serves to explain many diverse conditions found in the pathologic development of sex organs. Dorothy Price and her collaborators (see Price and Ortiz, 1965; Price *et al.*, 1967) have further elucidated the epigenetic control of sex organ development with the help of organ cultures. Their results show that the indifferent sex organs are only susceptible to the action of hormones for a limited period of time. Thus in guinea pigs, for example, the Wolffian ducts of female fetuses can be maintained if they are exposed to the androgenic action of the fetal testes, but this depends on the stage of development of the fetus. If an organ culture combination of the female genital tract and testes is made with organs that are too old, so that degeneration of the ducts has already started, androgens cannot prevent their degeneration. On the other hand, in the male fetal genital tract the maintenance of the same ducts is determined by the hormones quite early during development and subsequent removal of the testes does not lead to degeneration of the Wolffian ducts. The situation with the Müllerian ducts is analogous. If the Müllerian ducts have been induced to degenerate, removal of the testes cannot prevent the complemention of this process in the experimental situation. On the other hand, it seems that these ducts can also reach a stage of stability after which they are no longer susceptible to the destructive influence of male gonads.

The development of the external genitalia is under a similar control system. They develop in the male direction in the presence of large amounts of androgens and in their absence differentiate in the female direction. External organs, however, differentiate later than internal genitalia and, accordingly, they can be modified by hormonal treatment during late stages of pregnancy. Several types of intermediate forms may arise in pathologic conditions and can be interpreted in terms of the intensity and duration of the abnormal hormonal influence.

Effects of Other Hormones

Other hormones, such as adrenal cortical steroids, adrenaline, insulin, parathormone, and so on, would be expected to regulate the differentiation of several organs of the embryos but their effect on differentiation is not so well explored as in the above mentioned cases of amphibian metamorphosis and reproductive tract differentiation. Elaborate techniques are needed for the detection of hormonal effects in differentiation. Organ culture studies have been particularly useful in experimentation because development can be analyzed in the presence and absence of the hormones under study (see Fell, 1964). The morphologic and functional development of the mammary gland may serve here as an example of a model system illustrative of the possibilities of organ culture studies (Stockdale *et al.,* 1966; Lockwood *et al.,* 1967).

When the immature mammary gland starts differentiation in the pregnant mouse, two principles of differentiation can be distinguished:

1. Proliferation of the ductal epithelial cells with formation of alveolar secretory lobules.
2. Production of specific milk proteins such as casein.

To obtain full differentiation of the gland in organ culture according to the criteria above-mentioned, the presence of three hormones is required: insulin, cortisol (hydrocortisone), and prolactin. Omission of any one of these hormones leads to deficient development. By

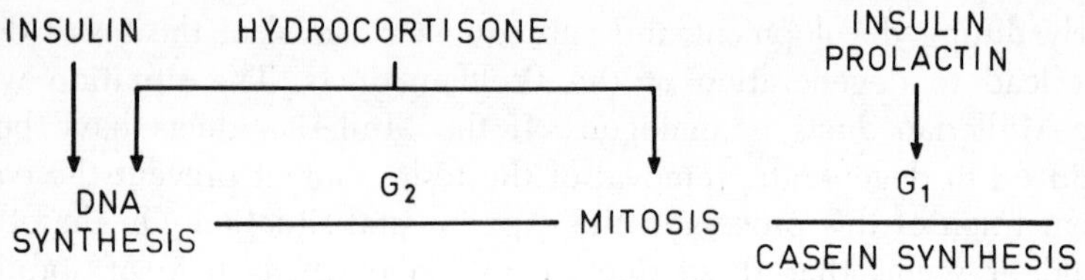

Fig. 6.3 Scheme of the hormonal requirements for mammary gland development in organ culture. (After D. H. Lockwood, F. E. Stockdale and Y. J. Topper. Science **156:** 945 [1967].)

omitting one or two of the hormones for some part of the incubation period and then adding them to the culture, a general scheme of the differentiation of the gland has been evolved (Fig. 6.3). Insulin alone is required for cell proliferation. However, cells proliferating in the presence of insulin but in the absence of cortisol cannot differentiate to casein-producing gland cells. Both prolactin and insulin are again needed for the specific synthesis of the postmitotic cells.

BIRTH DEFECTS CAUSED BY ENDOCRINOLOGIC FACTORS

The developing mammalian fetus may be exposed to an abnormal hormonal environment for a multitude of reasons; the extent to which its development will be affected depends on the nature, severity, and exposure time of the pathologic hormonal condition.

Reasons for the abnormally high hormone concentration in the fetal organism may be overproduction of hormones by fetal glands, overproduction of maternal hormones and their transfer to the fetal circulation, or exogenous hormone treatment of the mother. Deficiency of hormones may result from agenesis or dysgenesis of fetal glands, lack of some tropic hormone from the hypothalamus or hypophysis, some error in the biosynthesis of a hormone, or hormone-suppressant therapy applied to the pregnant mother. Some of these situations may be interrelated, so that hormone deficiency in one site leads to overproduction in another. This occurs, for example, in maternal diabetes, where the mother's insulin deficiency causes hyperglycemia in both mother and fetus and leads to hyperproduction of insulin in the fetal islets.

Examples of pathologic hormonal conditions and their consequences will be cited below without any attempt to treat this large field extensively.

Hypophysis

As mentioned earlier, the tropic action of the hypophysis on the peripheral endocrines is established during pregnancy. Therefore, it is somewhat surprising that the results of experimental intrauterine hypophysectomy do not indicate any major developmental failures except in the target endocrines. It is noteworthy that hypophysectomized embryos often reach the same size as normal fetuses at term. Because the surgical defects of the hypophyseoprivic fetuses are such that they cannot survive after birth, it is difficult to evaluate other defects caused by a missing hypophysis.

In the mouse, there are two types of pituitary dwarfism due to recessive genes. As a result of deficient hypophyseal hormone production, the mice do not grow and their endocrines are hypoplastic. Supplemental therapy with thyroxine and somatotropin repairs all the defects caused by the disorder in one of the mutant mice, but is less efficacious in the other type (Grüneberg, 1952; Bartke, 1963).

Pituitary dwarfism is also known in man and in some cases, at least, seems to be genetically determined. The retardation of growth becomes evident at about 3 to 4 years of age.

Thyroid Gland

Pathologic conditions related to abnormal thyroid development and endocrine function and their consequences are relatively common. Deficient thyroid function in an infant is relatively easy to recognize on the basis of the lethargy, dry skin, retarded growth, and typical laboratory findings. This is an important clinical problem because substitution therapy with thyroid hormones, if established early, restores all the somatic functions and may prevent the development of irreparable mental retardation, which is always associated with severe congenital hypothyroidism.

The etiology of congenital hypothyroidism may be divided into four main categories: defective development of the thyroid gland; prenatal treatment of the mother with ^{131}I or goitrogens (thiouracil or similarly acting substances); defective thyroxine biosynthesis; and nutritional deficiency of iodine (see Liu, 1966). The reasons for defective thyroid gland development are obscure. One familial case has been reported and it has been speculated that autoimmunity to maternal thyroid could prevent normal development of the thyroid in the offspring, but definitive proof is still lacking.

Treatment of the mother with thyroid inhibitors, thiouracil derivatives, and iodine causes fetal goiter in a high proportion of cases. These drugs block fetal thyroid hormone production with consequent overproduction of fetal TSH which then stimulates the growth of the thyroid gland. There are, however, some aspects of the fetal goitrogenesis subsequent to maternal drug treatment which are not understood. Burrow (1965) analyzed the pregnancy outcome of 41 gravidas who had taken propylthiouracil during gestation. Only five children of 37 liveborn had congenital goiter and these could not be related to the dosage, time of treatment, or thyroid status of the mother. It may be that the fetuses developing goiter had some subtle weakness in their capacity to synthesize thyroid hormones or that placental transfer of maternal thyroxine was inhibited to a greater extent than normally.

The treatment of maternal thyroid disease with radioactive iodine has a disastrous effect on the fetal thyroid. Iodine passes freely to the fetus and is concentrated in the functioning fetal thyroid, destroying it by virtue of the concentrated radioactivity. The damage caused by the radiation is irreversible.

The goiter and hypothyroidism caused by maternal treatment with iodine or thiouracil is normally transient. There is nevertheless a risk of irreversible damage, for the goiter may be so large that it mechanically causes respiratory difficulties or even suffocation, and somettimes the hypothyroidism of the infant may be so prolonged that irreparable mental retardation develops.

Defective thyroid hormone biosynthesis is one of the inborn errors of metabolism (see Chapter 3). Of the many steps in the biosynthesis of thyroxine, several may be defective—from the uptake of iodine by thyroid tissue to enzyme defects in the final steps of thyroxine synthesis.

Congenital hypothyroidism, with or without goiter, regardless of its etiology, has two main consequences: retarded growth as seen in cretinism and mental retardation. These disorders can be counteracted by thyroid therapy, but unfortunately the diagnosis is often made so late that the developmental arrest responsible for the mental deficiency is already irreparable.

Another congenital thyroid disorder sometimes encountered is congenital hyperthyroidism. If the mother has been treated with antithyroid drugs when pregnant, the removal of their influence after birth may lead to transient overproduction of thyroid hormones by the hyperplastic thyroid of the infant, causing a short period of hyperthyroidism. There is also another type of neonatal hyperthyroidism, which is longer-lasting, although likewise self-limited. A serum factor with long-acting thyroid-stimulating effect (LATS) has been described in patients with Graves' disease. This factor is different from TSH and has been shown to be closely associated with the serum

7S-gamma globulins, possessing many of the characteristics of this class of proteins (Kriss *et al.*, 1964). Some mothers with hyperthyroidism or with a history of hyperthyroidism have given birth to infants with transient congenital hyperthyroidism and LATS has been found in the serum of both the mothers and their infants (McKenzie, 1964). LATS graduallly disappears from the serum of the infants and there is a parallel disappearance of the hyperthyroid symptoms.

Does maternal thyroid disease cause congenital anomalies unrelated to the thyroid of the fetus? At present, this question is difficult to answer. Hoet *et al.* (1960) has suspected that maternal hypothyroidism may cause anomalies in human fetuses. Van Faassen (1957) could produce ocular anomalies in rat fetuses by performing thyroidectomy on the mother during gestation. The practical significance of hypothyroidism or hyperthyroidism as a cause of congenital anomalies in clinical medicine has still to be evaluated.

Congenital Disorders Related to Steroid Metabolism

As described on page 146 animal experiments have indicated that the androgenic hormones play a significant role in somatic sex differentiation. Thus it might be expected that their absence would lead to a feminization of genetic males, known as male pseudohermaphroditism, and that overproduction of androgens would cause a masculinization of genetic females, female pseudohermaphroditism. Both types can be produced experimentally in animals and are known in clinical medicine as well (see Jost, 1965).

The etiology of the abnormal hormonal environment of the fetus can be divided into three main categories:

1. Agenesis or dysgenesis of the gonads
2. Metabolic errors in steroid hormone biosynthesis
3. Hormonal imbalance (overproduction) in the mother or exogenous hormone treatment

Examples of each of these categories will be briefly described below as causes of deviant genital tract development.

Female pseudohermaphroditism is probably more common than its male counterpart. It is characterized by development of the external genitalia in the male direction, fusion of the labioscrotal folds, and hyperplasia of the genital tubercle to resemble a penis. In human female pseudohermaphrodites the internal genitalia are almost invariably derived from the Müllerian duct, that is, uterus and tubes are present. This is understandable in view of the local "inducing" effect of the testes as a cause of Müllerian duct degeneration (page 146). The female fetus may be exposed to very high concentrations of androgenic hormones, but they lack the ability to bring about degeneration of the Müllerian ducts.

Adrenogenital syndrome The best known cause and the most frequent type of female pseudohermaphroditism is congenital adrenal hyperplasia, the so-called adrenogenital syndrome. Detailed reviews of the topic have recently been presented by Bongiovanni and Root (1963) and Stempfel and Tomkins (1966). It is a hereditary disorder characterized by masculinization of the external genitalia of the newborn. The virilization progresses after birth and is accompanied by rapid growth during childhood as well as signs of the male type of pseudopuberty, such as deepening of the voice and development of facial hair. Growth ceases earlier than normal—at the age of 8 to 10 years—and untreated patients usually remain abnormally short. The degree of virilization may vary from very mild enlargement of the clitoris to external genitalia resembling those of a cryptorchid male. In the extreme cases female babies are likely to be raised as boys unless the underlying disorder is recognized. Regardless of the extent of the external virilism, the internal genitalia are always of normal female type and ovaries are present.

The pathogenesis of this disorder is due to excessive production of androgenic hormones secreted by the adrenal cortex. This, in turn, is a result of impaired cortisol biosynthesis, one of the main tasks of the healthy adrenal. The activity of some of the enzymes catalyzing cortisol synthesis is deficient and as a consequence steroid metabolism is channeled to an alternative pathway leading to androgenic hormones (see Fig. 6.4). As seen in this figure, some of the intermediate metabolites of the cortisol pathway are precursors of androgenic hormones and even under normal conditions small amounts of androgenic hormones (among them testosterone) are synthesized in the adrenal. Cortisol is the natural inhibitor of the ACTH production of the hypophysis and in its absence large amounts of ACTH are produced. This, in turn, leads to constant overproduction of the adrenal androgenic hormones.

In the most simple and common form of the adrenogenital syndrome, the defect is in the hydroxylation of the carbon at the 21-position. The block is not complete, but a faulty balance is established between cortisol and androgen production. As a result, virilization develops but the patients seldom suffer from adrenal insufficiency and sufficient amounts of aldosterone, the principal hormone regulat-

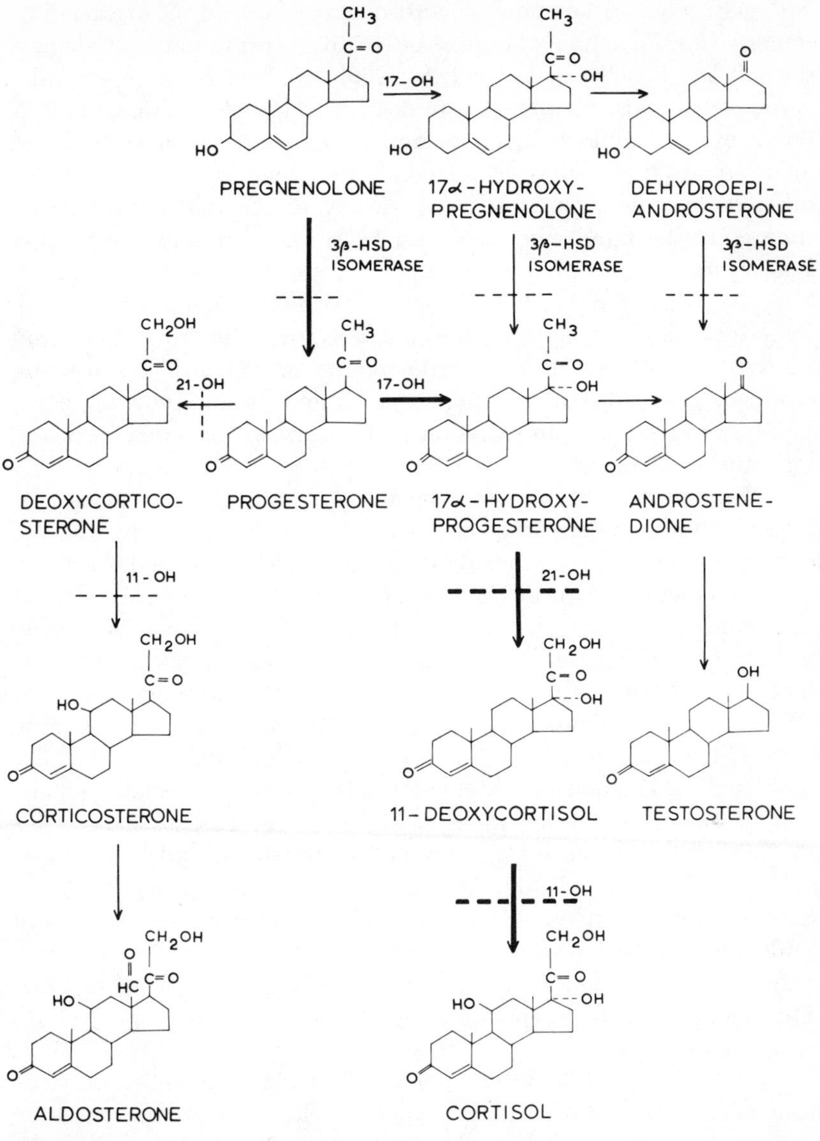

Fig. 6.4 Schematic representation of the biosynthesis of cortisol (thick arrows), aldosterone, and testosterone. The partial blocks caused by deficient enzymes are indicated with cross lines.

ing the water and salt balance, are produced. In a more severe form of the syndrome, in addition to the virilism, there are symptoms of severe adrenal insufficiency, with a tendency to lose salt from the body. In this form, the 21-hydroxylase activity is severely deficient and only minimal amounts of cortisol are produced, no matter how intensive ACTH stimulation may be. With regard to pathophysiology, the salt loss is believed to result from impaired synthesis of mineralocorticoid steroids, mainly aldosterone, as indicated in Fig. 6.4. It is not known whether 21-hydroxylation in the synthesis of cortisol and aldosterone is catalyzed by one and the same enzyme or by two different enzymes. The salt-losing variety of the adrenogenital syndrome may be fatal some days after birth unless appropriate therapy is instituted.

The third type of adrenogenital syndrome is associated with hypertension. In this form the block is in 11-hydroxylation, and 11-deoxycortisol and 11-deoxycorticosterone accumulate; no salt-losing tendency is found. Both the hypertension and the salt-retaining capability are attributable to secretion of 11-deoxycorticosterone, a potent mineralocorticoid.

The fourth type of adrenogenital syndrome is due to lack of 3β-hydroxysteroid dehydrogenase (HSD). In infants with this enzyme defect virilization of the female is less conspicuous or totally lacking. On the other hand, male infants often show underdevelopment of the external genitalia, such as hypospadias resulting from incomplete closure of the genital folds. This form of adrenogenital syndrome, first described in man by Bongiovanni (see Bongiovanni and Root, 1963), has been reproduced experimentally in rat fetuses by using a steroid analog, which inhibits HSD (Goldman *et al.*, 1966). The hormonal explanation of the slight virilization of the female and the underdevelopment of the male genitalia is that HSD is needed for the biosynthesis of all biologically active steroids, including testosterone (see Figs. 6.1 and 6.4). Accordingly, testosterone is not produced for the normal development of the male genital tract or for the full virilization of the females. The slight virilization in females, however, may be attributed to production of weak androgenic hormone metabolites, including dehydroepiandrosterone.

Failure of gonadal development There are a number of other congenital forms of deviating sexual development accompanied by abnormal hormone production. Many of them are associated with aberrations in the sex chromosomes (see Chapter 3). In some of them the following pathogenesis can be postulated. An abnormal sex chromosome pattern causes dysgenesis of gonadal development, which in turn leads to deficient sex hormone production and is reflected in abnormal differentiation of the genital tract. Interested readers are referred to the reviews of Jost (1965) and Federman (1967) on this subject. Only one, already classical, example will be briefly cited here.

In gonadal dysgenesis or Turner's syndrome, the most striking features are short stature and immature genitalia of female type with totally underdeveloped so-called streak gonads, in which no germinal elements can be found. These patients have a sex chromosome pattern of type XO. Their urine contains very little estrogens, but great amounts of gonadotropin indicating normal hypophyseal function. It has been postulated that these patients are human counterparts of rabbits experimentally castrated before differentiation of the sex organs has started (Federman, 1967).

Exogenous hormones Exposure of the fetus to an abnormal hormonal environment may occur not only as a result of endogenous causes but also due to maternal hormonal imbalance or hormone treatment. A rare condition has been described in which the pregnant mother had an androgen-producing ovarian neoplasm, arrhenoblastoma, and as a result the newborn female infant showed virilization of the genitalia. A more common and important source of anomalies is maternal hormone treatment. Testosterone treatment causes virilization of female fetuses, but some synthetic progesterone derivatives also have a virilizing effect on the fetus. These hormones have been used therapeutically in attempts to prevent threatening abortion and have caused virilization (Wilkins, 1960). This condition is usually less severe than the pseudohermaphroditism caused by the adrenogenital syndrome.

DIABETES MELLITUS

The most common endocrinologic disease of the mother which affects the development of the fetus is diabetes mellitus. Before the introduction of insulin therapy, overtly diabetic women seldom became pregnant or gave birth to a living infant. The situation has changed drastically and today diabetic mothers have good chances of uncomplicated pregnancy and successful delivery. The offspring of diabetic mothers, however, are often found to be suffering from pathologic conditions and perinatal mortality is considerably higher than among children of healthy mothers. The infant of a diabetic mother shows characteristic features in appearance, in development of internal organs, and in metabolism (see Kyle, 1963). These children are significantly heavier than normal, they are edematous and round-faced and have thick subcutaneous fat deposits (Fig. 6.5). They are often weak and have respiratory difficulties. Visceral organs such as heart, liver and adrenals are very large, but this visceromegaly is not uniform and some organs show retarded development. The kid-

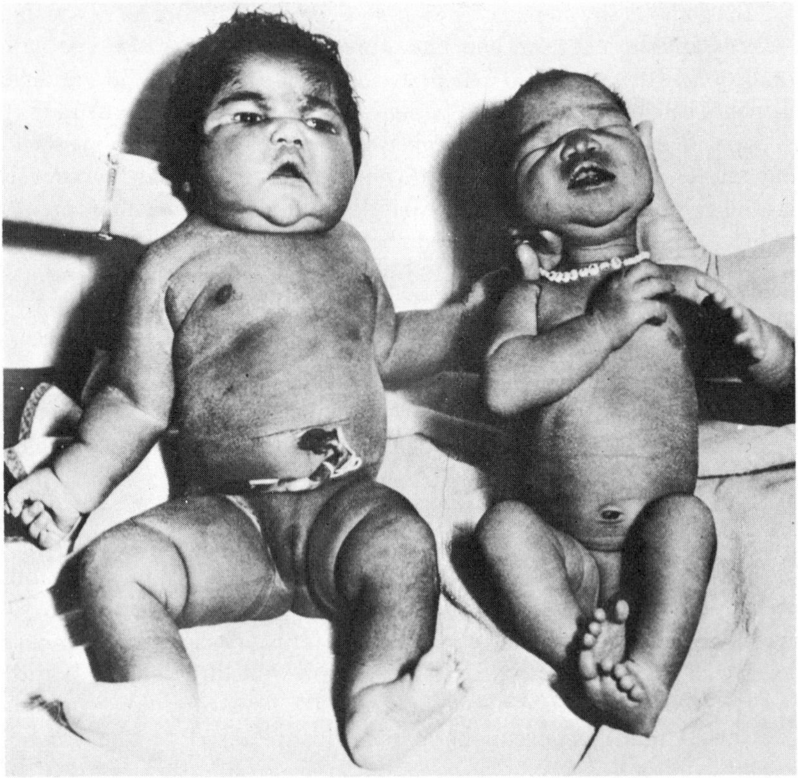

Fig. 6.5 A newborn child of a diabetic mother compared with a normal newborn. (From: W. P. U. Jackson, Lancet ii: 625, 1955).

neys are often smaller than normal and the renal glomeruli show retarded development. The brain is also diminished in size and the most conspicuous finding of retarded development is abundant extramedullary hematopoiesis at term. The islets of Langerhans are hyperplastic (Fig. 6.6).

Despite their large size and visceromegaly, these infants are less mature than their size would indicate.

Insulin activity in the cord blood is sometimes very high (Fig. 6.7) and the same has been reported for the fetal pancreas.

Children of diabetic mothers run a higher risk of perinatal death. There is a tendency to fetal death at week 37 of pregnancy and the total perinatal mortality varies from 10 to 40 percent in the different series reported, but is always higher than in the corresponding control material.

Since the first reports of increased incidence of congenital malformations in children of diabetic mothers, there has been much debate about the justification of this suspicion. Several recent and large studies show that it is warranted (Pedersen *et al.,* 1964). Examination of the malformed children has disclosed a wide array of anomalies, but recently a specific type of malformation has been described which is frequently found in the children of diabetic mothers, particularly if the diabetes has been severe. This type of malformation is characterized by defects in the lower ends of the femora (Fig. 6.8).

The metabolic etiology of the typical pathologic features of the infants of diabetic mothers are not fully understood. Most authors believe that all changes seen in the infant are due to the hyperglycemia transferred from mother to fetus. This would, in turn, cause

Fig. 6.6 Islets of Langerhans in the pancreas of a normal newborn, (top) and in that of a stillborn fetus of a diabetic mother (Courtesy of Dr. K. Krohn.)

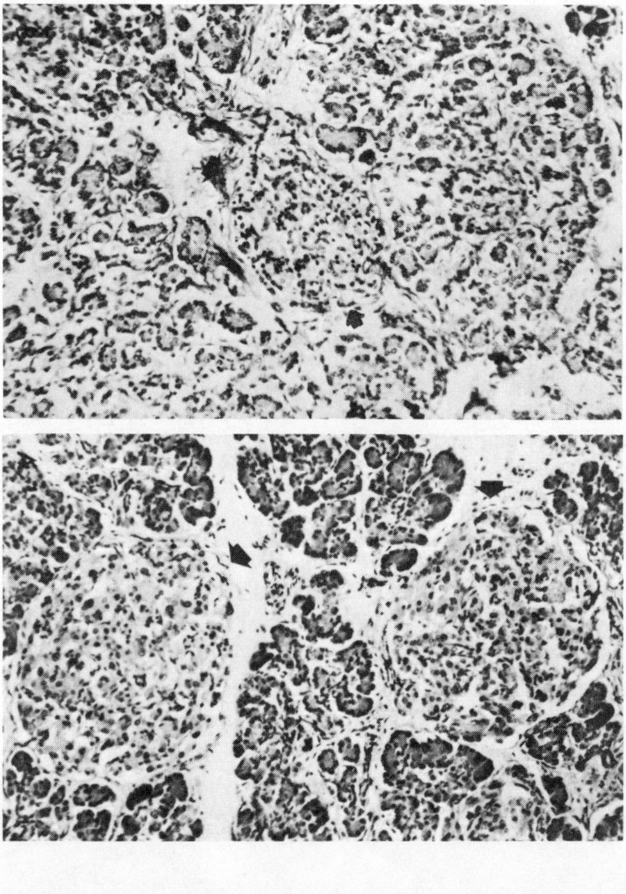

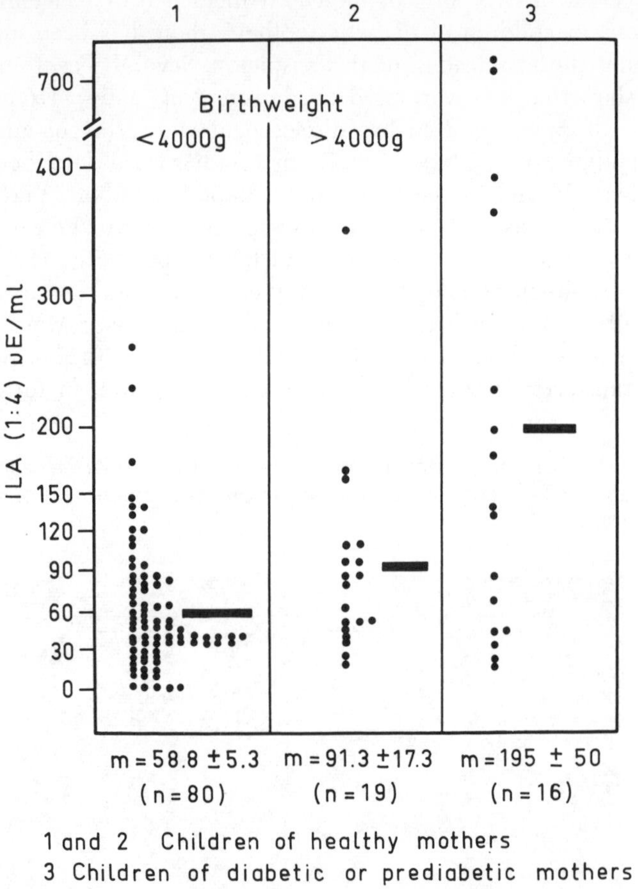

Fig. 6.7 Insulin-like activity (ILA) in the cord blood of children born to normal and diabetic mothers. Each dot represents one case and the horizontal bar, the mean value of ILA in each group. (After E. F. Pfeiffer and R. Ziegler. Triangle **7**: 8 [1965].)

overproduction of fetal insulin, all other changes being secondary to the hyperinsulinism of the fetus. Other hormonal and hereditary etiologic factors, such as overproduction of somatotropin, disturbances in adrenocortical hormone production, and so on, have been suspected, but never demonstrated (see Jackson, 1955; Kyle, 1963).

The typical heavy infants of diabetic mothers have been a stimulus for a new concept of diabetes mellitus. It was found that sometimes apparently healthy women gave birth to overweight children with all the anomalies typical of the children of diabetic mothers. It was discovered that these mothers developed overt diabetes some

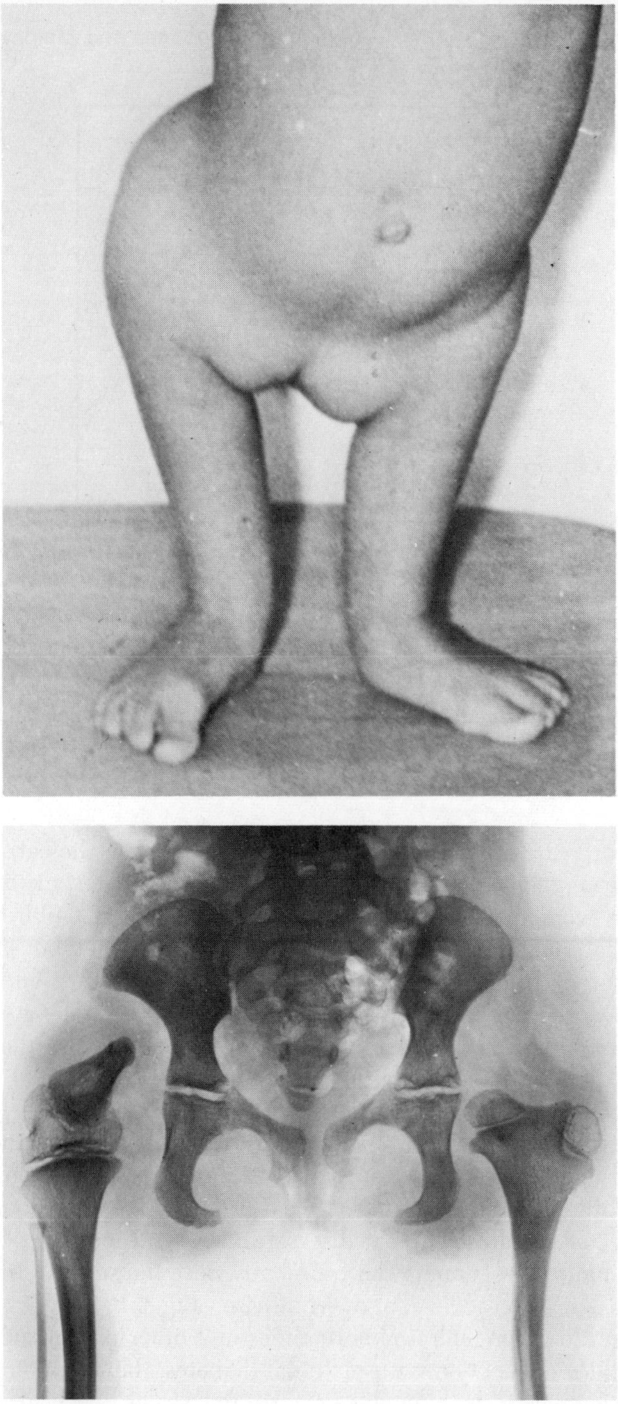

Fig. 6.8 Typically malformed child of a diabetic mother and roentgenograph showing defect in the right femur and hips. (From J. Kucera *et al* Dtsch. Med. Wschr. **90:** 901 [1965].)

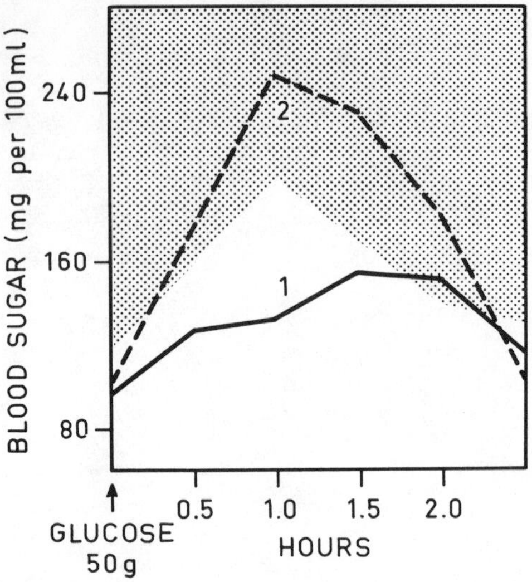

Fig. 6.9 Glucose tolerance curve of a mother who had given birth to a stillborn child weighing 5.7 kg. (1) Six days after the stillbirth; (2) two years later. White area represents normal glucose values and shaded area abnormal values after administration of glucose. (After W. P. U. Jackson. Lancet ii: 625 [1955].)

months or years after delivery (Miller, 1944). When the carbohydrate metabolism and other parameters related to diabetes mellitus were tested in women who had given birth to overweight children, it was found that in many tests they showed abnormal metabolism resembling that of diabetic patients (Jackson, 1955; Pfeiffer and Ziegler, 1965). On the other hand, it is evident that pregnancy itself is diabetogenic, that is, mothers with a predisposition to the disease showed diabetic metabolism during pregnancy or developed the overt disease after pregnancy (Fig. 6.9).

The present concept of this disease is that diabetes mellitus is an inherited disease present from birth and progressing slowly with age. Subtle variations in metabolism can be detected long before the onset of manifest clinical disease and several stresses of metabolism, including pregnancy, can transiently raise the subthreshold metabolic weakness to the level of overt disease (Fig. 6.10).

Regarding congenital abnormalities and other pathologic features described in connection with maternal diabetes, the question has been raised whether it is hypoglycemia or the drugs used for diabetes therapy that affect the development of the fetus. Here, the results of experimental studies with animals and clinical experience are somewhat contradictory, as is frequently the case in medicine. It has been

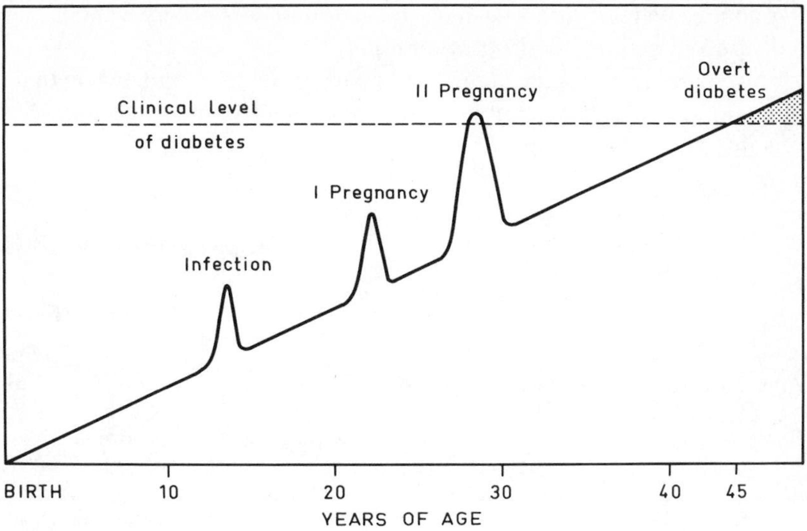

Fig. 6.10 Scheme of the course of diabetes mellitus.

repeatedly shown that high doses of insulin administered to pregnant laboratory animals may cause anomalies. Smitheberg and Runner (1963) studied the effect of insulin and other hypoglycemic treatments on several inbred strains of mice and noted that they cause anomalies of the skeleton and central nervous system, but the susceptibility of different strains to the hypoglycemic treatment varied. There is no evidence that insulin treatment of pregnant women increases the incidence of malformations in humans, but it is quite evident that careful treatment of the pregnant diabetic mother with insulin and other necessary therapeutic measures increase the chances of survival and normal development of the fetus (see Kyle, 1963).

HORMONES AS TERATOGENS

In this chapter and in some other contexts, it has been noted that exogenous hormones given to pregnant women may act as teratogens. It is true of many, if not all, known hormones that if given in large quantities to animals during the sensitive organogenetic period, they may cause abnormalities. In several instances, the mechanism of their teratogenic activity is not known. It seems that they work on a different metabolic basis than when exerting their physiologic function. This does not mean that they have no place in experimental teratology; we have shown on pages 20 and 82 that hormones such as cortisone have been found to be very valuable as tools for unravelling the laws of teratology.

In clinical medicine, however, apart from the virilizing activity of some steroids, the possible teratogenic effect of external hormones remains to be evaluated.

SUGGESTED READINGS

Review Articles

Allen, B. M. The endocrine control of amphibian metamorphosis. Biol. Rev. **13**: 1–19 (1938).

Bongiovanni, A. M. and A. W. Root. The adrenogenital syndrome. New Engl. J. Med. **268**: 1283–1289, 1342–1351, 1391–1399 (1963).

Etkin, W. Metamorphosis. *In* Analysis of Development, (B. H. Willier, P. A. Weiss and V. Hamburger, eds.) Sect. XII, pp. 631–663. W. B. Saunders Co., Philadelphia and London (1955).

Federman, D. D. Abnormal sexual development. W. B. Saunders Co., Philadelphia and London (1967).

Fell, H. B. The role of organ cultures in the study of vitamins and hormones. Vitamins and Hormones **22**: 81–127 (1964).

Gallien, L. G. Genetic control of sexual differentiation in vertebrates. *In* Organogenesis, (R. L. DeHaan and H. Ursprung, eds.), pp. 583–610. Holt, Rinehart and Winston, New York (1965).

Grüneberg, H. Endocrine organs. *In* H. Grüneberg: The Genetics of the Mouse, pp. 122–129. Nijhoff, The Hague (1952).

Hagerman, D. D. and C. A. Villee. Transport functions of the placenta. Physiol. Rev. **40**: 313–330 (1960).

Jackson, W. P. U. A concept of diabetes. Lancet II: 625–631 (1955).

Josimovich, J. B. Protein hormones and gestation. *In* Comparative Aspects of Reproductive Failure, (K. Benirschke, ed.), pp. 170–185. Springer-Verlag, New York (1967).

Jost, A. Problems of fetal endocrinology: the gonadal and hypophyseal hormones. Recent Progr. Hormone Res. **8**: 379–418 (1953).

——— Hormonal factors in the development of the fetus. Cold Spring Harbor Symp. Quant. Biol. **19**: 167–181 (1954).

——— The role of fetal hormones in prenatal development. Harvey Lectures, Ser. **55**: 201–226 (1961).

——— Gonadal hormones in the sex differentiation of the mammalian fetus. *In* Organogenesis, (R. L. DeHaan and H. Ursprung, eds.), pp. 611–628, Holt, Rinehart and Winston, New York (1965).

Klevit, H. D. Fetal-placental-maternal interrelations involving steroid hormones. Pediat. Clin. N. Am. **13**: 59–73 (1966).

Kyle, G. C. Diabetes and pregnancy. Ann. Internal Med. **59**: Suppl. 3, 1–82 (1963).

Liu, N. Some aspects of endocrinology in the fetus and the newborn. Ped. Clinics N. Am. **13**: 1047–1076 (1966).

Mitchell, F. L. Steroid metabolism in the fetoplacental unit and in early childhood. Vitamins and Hormones **25**: 191–269 (1967).

Price, D. and E. Ortiz. The role of fetal androgen in sex differentiation in mammals. *In* Organogenesis, (R. L. DeHaan and H. Ursprung, eds.), pp. 629–652. Holt, Rinehart and Winston, New York (1965).

Ryan, K. J. and L. Ainsworth. Comparative aspects of steroid hormones in reproduction. *In* Comparative Aspects of Reproductive Failure, (K. Benirschke, ed.), pp. 154–169, Springer-Verlag, New York (1967).

Stempfel, R. S. and G. M. Tomkins. Congenital virilizing adrenocortical hyperplasia. *In* The Metabolic Basis of Inherited Diseases, (J. B. Stanbury, J. B. Wyngaarden and D. S. Fredrickson, eds.), pp. 635–664. McGraw-Hill, New York (1966).

Wells, L. J. Fetal hormones and their role in organogenesis. *In* Organogenesis, (R. L. DeHaan and H. Ursprung, eds.), pp. 673–680. Holt, Rinehart and Winston, New York (1965).

Willier, B. H. Ontogeny of endocrine correlation. *In* Analysis of Development, (B. H. Willier, P. A. Weiss and V. Hamburger, eds.), pp. 574–619, W. B. Saunders Co., Philadelphia and London (1955).

Special Articles

Bartke, A. The response of two genetically different types of dwarf mouse to growth hormone. Genetics **48:** 882 (1963).

Burrow, G. N. Neonatal goiter after maternal propylthiouracil therapy. J. Clin. Endocrinol. Metab. **25:** 403–408 (1965).

Frandsen, V. A. and G. Stakemann. The clinical significance of oestriol estimations in late pregnancy. Acta Endocrinol. (Kbh) **44:** 183–195 (1963).

——— The site of production of oestogenic hormones in human pregnancy. III. Further observations on the hormone excretion in pregnancy with anencephalic foetus. Acta Endocrinol. (Kbh) **47:** 265–276 (1964).

French, F. S. and J. J. Van Wyk. Fetal hypothyroidism. J. Pediat. **64:** 589–600 (1964).

Goldman, A. S., A. M. Bongiovanni, and W. C. Yakovac. Production of congenital adrenal cortical hyperplasia, hypospadias and clitoral hypertrophy (adrenogenital syndrome) in rats by inactivation of 3β-hydroxysteroid dehydrogenase by (2α-cyano 4,4, 17α-trimethylandrost-5-en 17β-ol-3-one) administered to pregnant rats. Proc. Soc. Exp. Biol. Med. **121:** 757–766 (1966).

Hilfer, S. R. The stability of embryonic chick thyroid cells *in vitro* as judged by morphological and physiological criteria. Develop. Biol. **4:** 1–21 (1962).

Hoet, D. P., A. Gommers, and J. J. Hoet. Causes of congenital malformations: role of prediabetes and hypothyroidism. *In* Ciba Foundation Symposium on Congenital Malformations, (G. E. W. Wolstenholme and C. M. O'Connor, eds.), pp. 219–235. J. & A. Churchill Ltd., London (1960).

Jost, A. Problems of fetal endocrinology: the adrenal glands. Rec. Progr. Hormone Res. **22:** 541–574 (1966).

Kriss, J. P., V. Pleshakov, and J. R. Chien. Isolation and identification of the long-acting thyroid stimulator and its relation to hyperthyroidism and circumscribed pretibial myxedema. J. Clin. Endocrinol. Metab. **24:** 1005–1028 (1964).

McKenzie, J. M. Delayed thyroid response to serum from thyrotoxic patients. Endocrinology **62**: 865–868 (1958).

Miller, H. C. Cardiac hypertrophy and extramedullary erythropoiesis in newborn infants of prediabetic mothers. Am. J. Med. Sci. **209**: 447–455 (1944).

Pedersen, L. M., I. Tygstrup, and J. Pedersen. Congenital malformations in newborn infants of diabetic women Lancet **I**: 1124–1126 (1964).

Price, D., E. Ortiz, and J. Zaaijer. Organ culture studies of hormone secretion in endocrine glands of fetal guinea pigs. III. The relation of testicular hormone to sex differentiations of the reproductive ducts. Anat. Record **157**: 27–41 (1967).

Smithberg, M. and M. N. Runner. Teratogenic effects of hypoglycemic treatments in inbred strains of mice. Am. J. Anat. **113**: 479–489 (1963).

Stockdale, F. E., W. G. Juergens, and Y. J. Topper. A histological and biochemical study of hormone-dependent differentiation of mammary gland tissue in vitro. Develop. Biol. **13**: 266–281 (1966).

Van Faassen, F. Hypothyreoidie en aangeboren misvormingen. Oosterbaan & Le Cointre N.V., Goes (1957).

Wallace, S. J. and E. A. Michie. A follow-up study of infants born to mothers with low oestriol excretion during pregnancy. Lancet **II**: 560–563 (1966).

Weber, R. Biochemistry of amphibian metamorphosis. *In* The Biochemistry of Animal Development, (R. Weber, ed.) pp. 227–301. Academic Press, New York (1967).

Weiss, P. and F. Rossetti. Growth responses of opposite sign among different neuron types exposed to thyroid hormone. Proc. Nat. Acad. Sci. (U.S.) **37**: 540–556 (1951).

Wilkins, L. Masculinization of female fetus due to use of orally given progestins. J. Am. Med. Assoc. **172**: 1028–1032 (1960).

7

Radiation Hazards

The discovery of x-rays by W. K. Röntgen and the realization of their great possibilities in many fields of human activity were soon followed by the observation that radiation has dangerous effects on living cells and tissues and that its uncontrolled use would be a serious threat to health. Later research has disclosed that all ionizing radiations whether electromagnetic like x-rays and γ-rays or particulate like the alpha- and beta-radiation emitted from radioactive isotopes, have similar pathologic effects.

Ionizing radiation influences biologic systems in two different ways. The effect of high levels of radiation becomes apparent in cell death or severe injury soon after irradiation, whereas low levels may cause long-term effects such as mutations and cancerogenesis which may not appear until months or years after exposure.

The dose is not the only factor determining the effect of radiation. The various tissues and cells differ extensively in their sensitivity to the effects or radiation. It is well known that cells undergoing rapid proliferation, such as those in the bone marrow of the adult organism and cells in many parts of the embryo, are particularly susceptible to the injurious action of radiation.

It is believed that the physical basis of the tissue damage inflicted by radiation is the ionization of the atoms encountered by x-rays or high energy particles. This causes highly reactive radicals in the important cell constitutents themselves or in mediating particles such as the dissociation products of water. The penetration and quantitative effect per energy unit varies with the different types of high energy radiation. The pathologic effect, however, as seen in tissue damage and prevention of cell proliferation, is surprisingly uniform with all types of ionizing radiation, although small variations have been detected. Hitherto, most research in radiobiology has been performed with the x-rays or γ-rays, but the importance of radioactive isotopes is increasing.

In the following discussion two units of radiation will be used. The röntgen (r) is the unit for measuring the x-radiation or γ-radiation dose as delivered from the radiation source or when it meets the target. The *rad* is the amount of radiation energy absorbed by biologic material. The energy absorbed by one gram of ordinary soft tissue on exposure to 1 r of x-rays is roughly equal to one *rad*. Standard textbooks on biologic radiology, such as the one by Bacq and Alexander (1961), give a more detailed description of the physical characteristics of different types of radiation and their biologic effects.

All biologic aspects of radiation are relevant in comprehensive discussions about embryonic radiobiology. Here, however, we will limit the discussion to the more immediate consequences of radiation causing embryonic death or anomalous embryonic development apparent right after birth.

RADIATION INJURY ON THE CELLULAR LEVEL

There has been much discussion about the initial (chemical) lesion caused by ionizing radiation in cells. Its nature has not yet been discovered, but DNA, RNA, RNA-protein complexes, cell membranes, and related organelles have been regarded as the primary targets of the radiation injury. It is quite evident that the nucleus and cell-replicating machinery are the most vulnerable parts of the cell; cell division is inhibited with relatively low doses of radiation, the effect being most conspicuous in cell populations undergoing rapid proliferation. It is also well established that irradiation causes damage to the chromatin material of the cell nucleus. Furthermore, current knowledge of gene structure and function, on the one hand, and the well known mutagenic effect of radiation, on the other, suggest that genomic DNA is the main, if not the only important target of radiation.

Reproductive Death Following Irradiation

The techniques of cell culture have made possible studies on the effects of radiation on animal cells. When rapidly proliferating cells in cultures are irradiated with different doses, a proportion of the cells lose the ability to proliferate and eventually die, a phenomenon known as reproductive death (Puck, 1964). This is illustrated

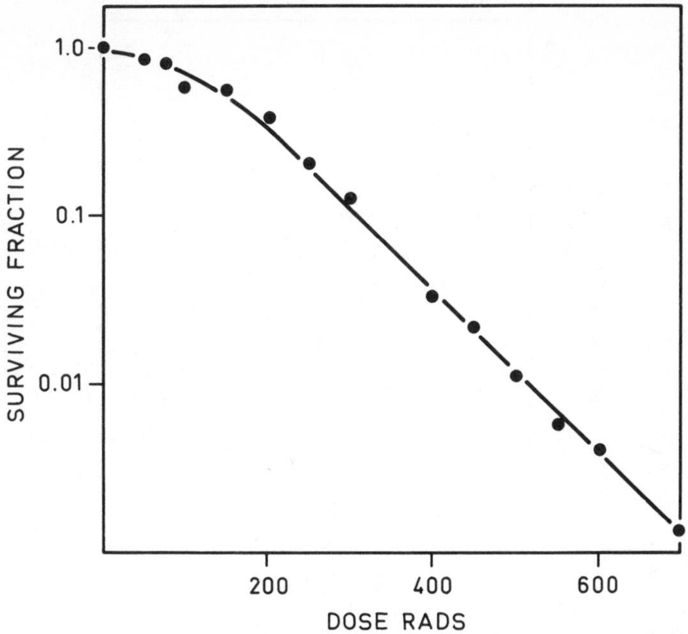

Fig. 7.1 Effect of increasing dose of x-radiation on the survival of HeLa cells *in vitro*. (After T. T. Puck. Cytogenetics of Cells in Culture. R. J. C. Harris [ed.], p. 63. Academic Press, New York and London [1964].)

in an exponentially growing HeLa cell culture in Fig. 7.1. The killing action (or survival) is related to the dose of radiation given. When the dose of radiation needed to reduce the fraction of surviving cells to 37 percent (D_0) is calculated, (it is 128*r* in this particular experiment) it has been found that all cells derived from warm-blooded animals have similar D_0 within a factor of 2 or 3.

Relation Between Stage in the Cellular Life Cycle and Reaction to Radiation

Cell culture studies have also shed light on the question of whether or not the differential radiosensitivity of cells depends on the stage at which they are irradiated. The life cycle of a cell is commonly divided into four main stages: mitosis (M) is the time of actual replication of the nucleus and cell; this is followed by a gap (G_1); then the nuclear DNA is synthesized (S); after the S period there is a second gap (G_2) before the cell embarks on a new mitosis. With various methods it is possible to induce all cells in a culture to follow this scheme simultaneously, resulting in so-called syn-

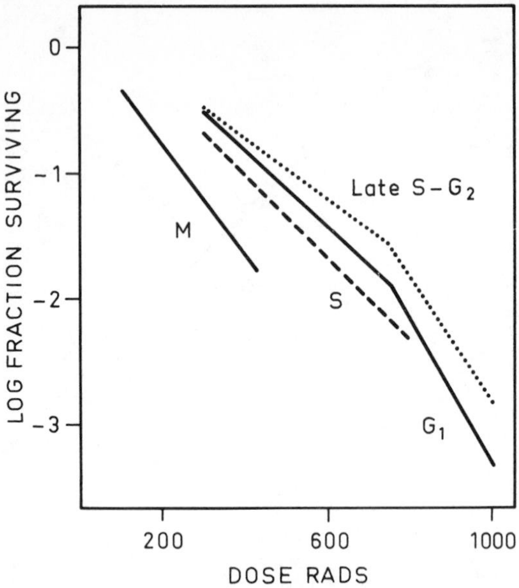

Fig. 7.2 Survival curves of samples of a synchronous HeLa cell culture irradiated with increasing doses of x-rays at different stages of the life cycle. (After L. J. Tolmach, T. Terasima and R. A. Phillips. Cellular Radiation Biology, p. 376. Williams and Wilkins, Baltimore [1965].)

chronous culture. When synchronous cultures are irradiated, the effect on survival varies with the stage of the cell in the cycle. This is illustrated in Fig. 7.2, where a synchonous cell culture is irradiated with increasing doses of x-rays and survival is calculated in relation to the stage of the cycle at the time of irradiation.

In the experiment in Fig. 7.2, cells undergoing mitosis suffer most in their reproductive capacity, the late S or early G_2 phase being the most resistant. The differential sensitivity to the cytocidal action during different stages of the cell cycle has been shown in several other experiments with animal and plant cells, but opinions differ about the relative sensitivity of the different stages, depending on the design of the experiment and cells used in the studies.

Radiation also affects the cells, which are not killed but continue their proliferation in culture. However, the proliferation is retarded as a result of radiation. This effect, too, is dependent on the stage at which the cells are in the time of irradiation. In Fig. 7.3 a summary of such experiments is shown. It seems that the delay in division is smallest after irradiation in early G_1 stage and increases as the cycle proceeds.

Further studies have shown that the proliferation-delaying effect

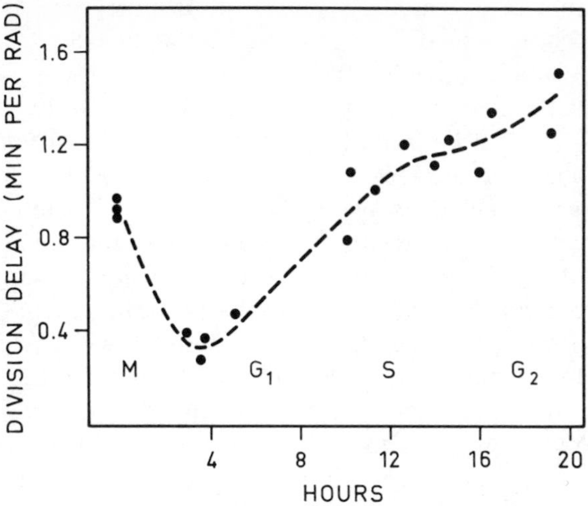

Fig. 7.3 Delay in cell division measured in populations of cells x-irradiated at the times after mitosis indicated on the abscissa. Approximate cell stages are indicated in the figure. (After L. J. Tolmach, T. Terasima and R. A. Phillips, Cellular Radiation Biology, p. 376. Williams and Wilkins, Baltimore [1965].)

may be due to a decrease in the rate of synthesis of DNA with prolongation of the S phase (Tolmach *et al.*, 1965).

Effect of Radiation on the Chromosomes

A long known visible effect of ionizing radiation on the cells is damage to the chromosomes. Chromosomes in cells are ordinarily studied microscopically at the metaphase or sometimes at the anaphase of mitosis. As a result of irradiation the chromosomes at metaphase appear to be "sticky" or clumped and the sharp boundaries between them are lost. At anaphase incomplete separation of chromatids is accompanied by bridges of chromatin material between the cells. This may lead to incomplete division of the chromosome material in the daughter cells. Another well-known effect of radiation on chromosomes is the breakage of chromatids or whole chromosomes. The broken fragments may be deleted from the chromosomes or they may rejoin in the original chromatid or in the other broken chromatids, causing several kinds of abnormal chromatin structures, such as ring chromosomes, multicentric chromosomes, and so on (see Chapter 3, page 60). Again, in chromosome breakage, the stage of the cell cycle at the time of irradiation determines the extent and nature of the chromosomal damage. Several studies on plant and animal cells have indicated that the G_2 stage of the life cycle is the most sensitive to chromosomal damage caused by x-rays (Scott and Evans, 1967; Dewey and Humphrey, 1965).

Some of the results regarding the differential sensitivity of tissue culture cells seem to be inconsistent and, in fact, not all reports agree about the phase of the cell cycle most sensitive to injurious effects of irradiation. As noted on page 168 and in Fig. 7.2 and 7.3, HeLa cells grown *in vitro* are relatively insensitive to the lethal action of irradiation at the G_2 phase and more recent results obtained by the same authors tend to confirm these results (Djordjevich and Tolmach, 1967). On the other hand, G_2 is the phase at which chromosome damage is found most frequently and some authors, using other cell lines, have postulated that this phase is also sensitive to the lethal action of irradiation (Sinclair and Morton, 1963). These inconsistencies may be due to differences in experimental design and in calculations of the results. Differences in the handling of cells and time of measuring the results may lead to conflicting conclusions, as has been shown by Evans (1968), when dealing with the frequency of chromosome damage in human leukocytes grown *in vitro*. It is reasonable to expect, however, that these lines of experimentation will reveal some of the basic mechanisms by which irradiation injures embryonic and adult cells *in vivo*. Some evidence of the differential susceptibility of the cleaving embryos is presented on page 174.

Other Factors Influencing the Radiosensitivity of Cells

When cells and chromosomes are injured by ionizing rays, they try to repair the damage. The mechanisms of the repair process are not fully understood, but it is possible that the differential radiosensitivity in various phases of the life cycle is as much a consequence of the varying recuperative powers in relation to the stage of the cycle as of actual differential sensitivity. The broken ends of the chromatids will rejoin after irradiation, but the repair process requires energy and protein synthesis. Therefore, procedures which interfere with synthesis of protein or energy transfer will lessen the extent of repair and thus intensify the chromosomal radiation damage (Wolff and Luippold, 1955; Wolff, 1967). Repair of the damaged chromosomal DNA apparently also requires DNA synthesis and it is to be expected that factors inhibiting it might augment the nuclear damage caused by radiation. It has been shown that some nucleic acid precursor analogs, such as fluorodeoxyuridine, or other inhibitors, such as hydroxyurea, render the cells more vulnerable to the action of radiation (Kaplan *et al.*, 1962; Weiss and Tolmach, 1967). On the other hand, certain treatments of tissue culture cells, such as hypoxia, low temperature, and compounds having free sulfhydryl groups (cysteine, mercaptoethanol, and so on) are known to reduce the injurious effects of radiation. The protective mechanism is largely unknown, but possible mechanisms are discussed in recent volumes of radiobiology (for example, *Cellular Radiobiology,* 1965). Recently, the existence of a specific enzyme or enzymes capable of rejoining broken ends of DNA strands of bacteria has been proposed and it is possible that similar mechanisms are concerned in the repair of the radiation-induced damage in the cells of higher animals as well (Gefter *et al.,* 1967).

THE EFFECT OF IRRADIATION ON THE EMBRYO

The embryo is a mosaic of cells of different kinds in different stages of development and comprises groups of cells proliferating at different rates. Even when the immediate effects of radiation, as revealed by a follow-up study of cell destruction, are relatively small, the consequences, for example, malformations, may be very severe. It must be remembered that, for normal embryonic development, proliferation, migration, and interactions of the cells have to occur in a highly organized and synchronized manner. Thus it is understandable that inhibition of cell multiplication or even a short delay at a critical point of development may have far-reaching consequences in disturbing the normal pattern of differentiation.

Lethal Action of Radiation

Radiation, if sufficiently intensive, can kill the embryo, but the sensitivity fluctuates during gestation. A study of the lethal dose at the different stages of mouse embryo development is presented in Fig. 7.4.

It is evident that the embryos are more sensitive to the killing action of x-rays in their early development, before the onset of major organogenesis, but even during these days the lethal dose varies by a factor of three or four. The radioresistance then increases steeply after organogenesis. It is noteworthy that the 17-day-old fetus is at least as radioresistant as its mother; it can hardly be killed *in utero* without killing the mother. It is probable that the action of x-rays on the embryo is a direct one and that indirect effects through maternal metabolism or the placenta, if they exist at all, are minimal.

Teratogenic Action

The teratogenic effect of ionizing radiation is the second major function of prenatal exposure. It was only 12 years after the discovery

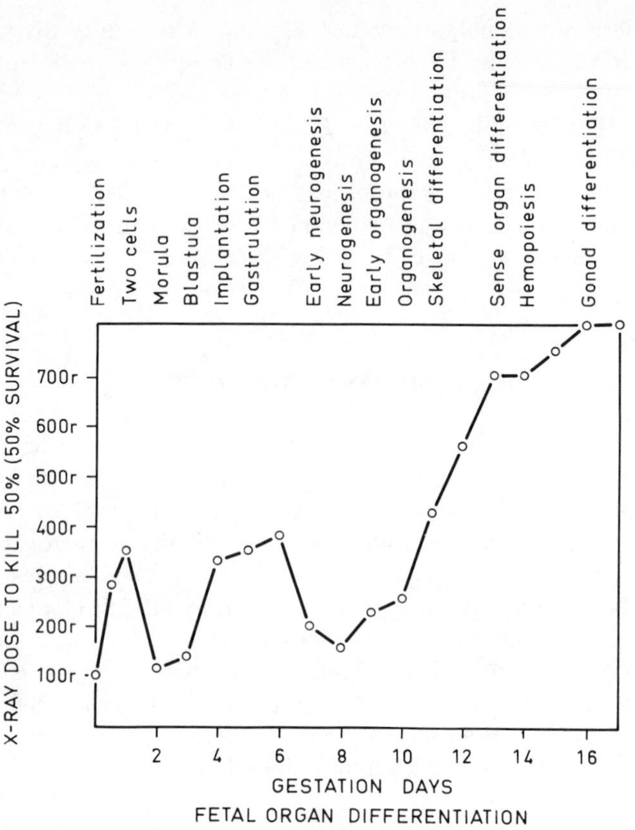

Fig. 7.4 The stage of pregnancy in days compared with the lethal action of x-rays in mouse embryos. The approximate stage of embryonic development each day is indicated in the upper part of the figure. (After R. Rugh and M. Wohlfromm J. Exp. Zool. **151:** 227 [1962].)

of x-rays that Hippel and Pagenstecher, in 1907, found that newborn rabbits had microphthalmia, cataract, and defective eyelids as a result of maternal x-ray irradiation during the early stages of pregnancy. Since then, ionizing radiation has been one of the main tools used in experimental teratology. The teratogenic effects of prenatal irradiation in experimental animals have been reviewed by a number of authors (Hicks, 1953; Russel, 1954; Russel and Russel, 1954; Hicks and D'Amato, 1966; and many others).

The concept of a sensitive period for congenital defects has largely developed as a result of irradiation experiments (see Chapter 5 and Fig. 5.9). Irradiation experiments have also shown that there is no sensitive period for the whole organism, but that each organ or tissue is sensitive to the teratogenic action when it is in active proliferation or in process of cellular differentiation. In most organs, this occurs during the relatively sharply limited organogenetic period during the first trimester of embryonic life. Similar but more limited differentiation continues in some organs during the latter half of prenatal life, a time which is considered resistant to the teratogenic action of most environmental factors. Ionizing radiation, however, is capable of causing malformations in fetuses after the organogenetic period, but only in those tissues which are still in process of differentiation, that is, mainly in the central nervous system and particularly in the cerebellar structures. The scheme showing sensitivity of the central nervous system to the teratogenic action of x-rays at different times of rat gestation is presented in Fig. 7.5.

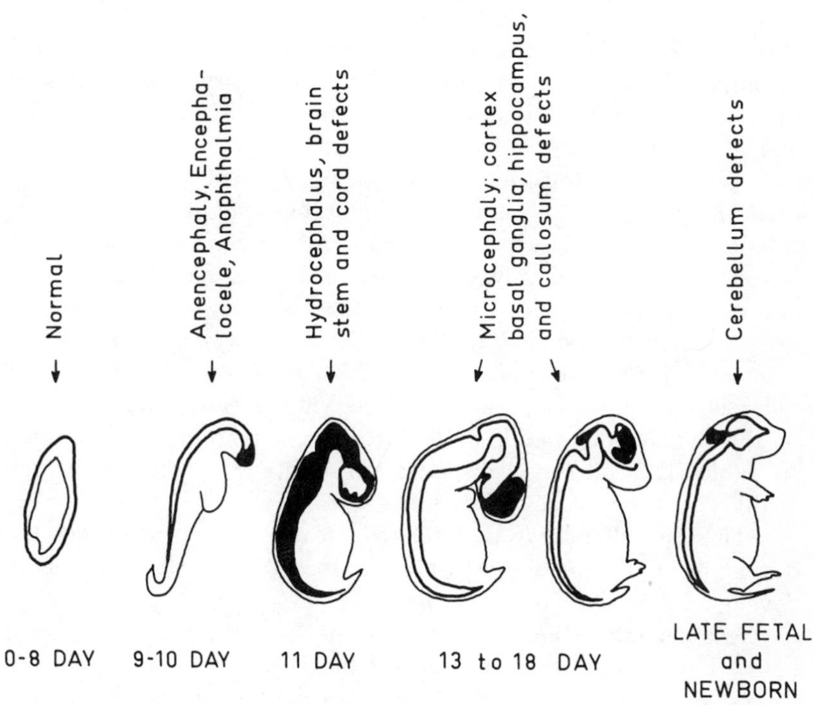

Fig. 7.5 Schematic drawing of central nervous system malformations caused by radiation in the rat at different times of gestation. (From S. P. Hicks, J. Cell. Comp. Physiol. **43** Suppl. 151 [1954].)

The number and severity of the malformations produced by irradiation is related to the intensity of the exposure. Larger doses of radiation produce more severe malformations in a higher proportion. Thus it has been shown that x-irradiation with 12.5 *r* only occasionally causes structural anomalies in rat offspring. Doses of 12.5–50 *r*, when given at the most sensitive stage of development, produce eye anomalies, central nervous system anomalies, and minor skeletal anomalies. Doses of 100–200 *r* produce a wide array of malformations in almost all the organ systems of rats and mice (see Brent, 1960).

Effect of Radiation on the Early Zygote

The largest number of congenital defects are undoubtedly induced during the organogenetic period, although hindbrain anomalies can be produced much later in development. It is believed that early development, that is, from fertilization to the beginning of organ formation, is the time when the embryo will either be killed or will develop normally. We have indicated in other chapters that there may be some exceptions to this rule. Recent experiments with x-irradiation elucidate the response of an early embryo and also show that sensitivity during the early cleavage stages will vary within very short time limits. L. B. Russell and her associates (see Russell, 1965; Russell and Montgomery, 1966) have performed irradiation experiments on mice shortly after fertilization and during the first and second days of development. They found two main consequences of irradiation, a lethal effect and chromosomal aberrations, which may be causally linked. In rigidly controlled experiments it was found that on the day of fertilization (day 0) and on the following day (day 1) the early embryos responded to the radiation in a strikingly different manner, depending on the hour at which they were irradiated. This difference is apparently due to the varying sensitivity and depends on the stage of the mitotic cycle, just as in cells irradiated in culture. In Table 7.1 is presented the radiation sensitivity of the early zygote at different times of development.

The injured embryos died relatively early in their post-irradiation development. Those embryos which escaped early death subsequently developed as well as control embryos.

Another interesting observation came from these same investigations. It was found that exposure of the cleaving embryos to x-rays causes loss of the sex chromosomes. A mouse with the sex chromosomal constitution X/O is phenotypically a normal female with normal mating capacity (O/Y is lethal). These embryos can be recognized phenotypically by using appropriate sex-linked markers. It was found

TABLE 7.1
Radiation Sensitivity of Early Mouse Embryos at Different Stages of Early Development.
The criteria for sensitivity are embryonic death and sex chromosome loss.[a]

Time	Stage	Radiation Sensitivity
Day 0 8:30 am	Most zygotes completing second-polar-body formation; less than one-quarter already in early pronuclear stage	+++++
Day 0 11:00 am	Most zygotes in early pronuclear stage	+++++
Day 0 3:00 pm	All zygotes in pronuclear stage; pronuclear growth and DNA synthesis probably completed	+
Day 1 Mid am	Almost all embryos in 2-cell stage, interphase	++
Day 1 3:30 pm	Most embryos beginning second cleavage	+++++

[a] From L. B. Russell and C. S. Montgomery: Int. J. Radiat. Biol. **10:** 151, 1966

that the susceptibility to X-chromosome loss was paralleled by that of the mortality when irradiation was performed at different times on the first day of development (Table 7.2).

TABLE 7.2
Frequency of Sex Chromosome Loss after Irradiation at Various Times of Day 0.[a]

Time Irradiated	Total Offspring Classified	Percent Frequency of Loss of		
		X^P or Y	$X^{M b}$	Total
Control	196	0.51	0	0.51
8:30 am	201	1.00	2.98	3.98
11:00 am	227	1.76	1.76	3.52
3:30 pm	193	0.52	0	0.52

[a] After L. B. Russell and C. S. Montgomery: Int. J. Radiat. Biol. **10:** 151, 1966
[b] Estimated frequency of occurrence, calculated by taking account of the fact that only half of all losses of X^M are detectable (since O/Y is an early lethal).

From Table 7.2 it is evident that there is a strong contrast between the radiation effects at different times during the same day.

Even more surprisingly, the data show that embryos irradiated during the later stages of the second meiotic division (irradiated at 8:30 AM) lose the maternal X chromosome almost three times as often as the paternal chromosome.

The coincidence of both lethal effect and x-chromosome loss at the same stages of development imply a common mechanism for both. Some other observations by the same authors and evidence from other sources support the view that loss of chromosomes is the basic cause of embryonic death. When the cleaving embryos are analyzed histologically on the days following irradiation, extranuclear bodies of chromatin are frequently found. X-chromosome losses in the survivors suggest the possibility that autosomes can also be lost as a result of irradiation. There are good reasons for assuming, however, that autosomal loss is lethal.

Mathematical calculations based on the known frequency of loss of the X-chromosome and the frequency of embryonic mortality, supposing that irradiation causes autosomal loss as often as sex chromosome loss and that autosomal monosomy is invariably lethal, give results that tally closely with the actual survival rate observed. Thus it may be that the lethal effect of x-rays on the zygote is due to chromosomal deletions, the loss of one sex chromosome being the only aberration not incompatible with survival.

Whether irradiation at early stages of development can cause actual structural malformations is still a controversial matter. Such malformations have been reported in the series of some authors, but their incidence is low and hardly differs from that of unirradiated controls.

Effect of Prenatal Radiation on Behavior

Irradiation experiments have also disclosed other aspects of embryonic pathology which are studied much less than mortality and teratology. It is known, for example, that irradiation can cause marked intrauterine growth retardation, the most sensitive period being shortly before the most sensitive period for the teratogenic effect in rats (see Brent and Jensh, 1967).

A line of investigation of the late effect of teratogenic treatment pioneered with the aid of intrauterine irradiation have been the studies concerned with the effect of radiation on later behavior and intellectual capacity. As shown previously, radiation can cause structural malformations in the central nervous system. It is of interest to try to relate these anatomic defects to central nervous system function and possibly to detect some specific functional deficiencies typical of a given type of morphologic injury.

Several animal experiments have been performed in which animals were irradiated at different times of gestation or postnatally and the function of the central nervous system subsequently registered by recording electrical brain rhythms or by using several tests of learning and other animal "intellectual" capacities. Such experiments (see Hicks and D'Amato 1966) have indicated that irradiations of the central nervous system during embryonic development can cause impaired brain function. Irradiation in the later stages (days 17 to term) of intrauterine development of rats result in offspring with abnormal electrical brain rhythms. Kaplan *et al.* (1963) have irradiated rats with relatively low doses (50 *r*) of x-rays at different times of development and postulate that there is a sensitive period for the induction of impaired learning shortly before the main central nervous system organogenesis (7½ days of gestation) and soon after the main period of anatomic formation of the brain (12 days and later). Other authors have noticed that the correlation between the anatomic lesion and the learning capacity of the irradiated animals may be very complex. For example, rats with very small cerebral hemispheres and defective cerebral cortices, consequent on 200 *r* radiation on day 17 of pregnancy, responded to a signal to obtain food as skillfully as nonirradiated rats; but if the food was only available for a limited period of time each day they, in contrast to the normal rats, did not have adequate motivation to get sufficient food during this period to maintain their body weight (see Hicks and D'Amato 1966).

EFFECT OF RADIATION ON HUMAN DEVELOPMENT

In the early years of medical radiology, when the deleterious effects of radiation were not fully realized and the standards for radiation protection were less rigorous, human embryos were exposed to radiation relatively often. Most of the information concerning the teratologic effects of radiation in man dates from this era. The best known study is that of Goldstein and Murphy (1929) based on a survey of the outcome of 106 pregnancies during which therapeutic radiation was given. Of the 75 children born alive, 38 were reported to be unhealthy. In 10 cases the radiation was considered not to have caused the congenital illness. Among the remaining children 14 microcephalics, 2 hydrocephalics, 1 mongoloid, and 3 with skeletal malformations were reported. In most of the microcephalic children the radiation was estimated to have occurred before the fifth month of gestation. There are a number of later reports of possible radiation effects on human fetuses, each of them based on observations on one or a few cases. In cases where radiation injury seems probable, microcephaly and postnatal mental retardation have frequently been reported. A limited number of autopsy reports agree with the observations made in animal experiments (see Hicks and D'Amato, 1966).

After the dangers of therapeutic irradiation during pregnancy were realized, the data collected from the Japanese atom bomb survivors in Hiroshima and Nagasaki added further information on the immediate effects of ionizing radiation on the human embryo. Yamazaki *et al.* (1954) reported increased fetal, neonatal, and infant mortality in offspring of pregnant mothers heavily irradiated during the explosion. The other significant finding of these authors and others (Plummer, 1952; Miller, 1956) was increase of the frequency of microcephaly and mental retardation, most commonly encountered in children exposed to radiation between weeks 7 and 15 of gestation. Anthropometric measurements also showed significant growth retardation in the children most heavily irradiated. These reports again agree with those known from animal irradiation experiments.

Other Sources of Ionizing Radiation and the Effect of Low Level Irradiation during Intrauterine Life

There are a number of other sources of ionizing radiation in addition to those reviewed above. Radiation from the fallout from nuclear detonations, diagnostic x-ray examinations, and radioactive isotopes used in the medical services and in research are all potentially hazardous to development. It has also been claimed that under some conditions cosmic and terrestial radiation may be dangerous. It is also likely that with the increased use of nuclear energy, isotopes, and other radiation sources in many fields of contemporary life, accidents will occur in which people will be exposed to large doses of ionizing radiation.

Malformations, significant growth retardation, mental retardation, and perinatal death are expected to result only from high levels of irradiation. The question of the possible danger of diagnostic medical x-ray investigations is still open to debate. It may be stated that with modern technology, by which unnecessarily large doses are avoided, the risk of a malformed child as a result of diagnostic radiation is slight—definitely less than the risk of omitting an important examination with strong indications. However, a good policy in daily x-ray routine is to restrict examinations of fertile women to the first ten days of the menstrual cycle in order to minimize the risk of irradiating an early embryo. Radioactive isotopes may be particularly dangerous because they increase the local tissue radiation to a very high level. Some isotopes, such as radioactive iodine and strontium, concentrate selectively in the thyroid and bones, respectively, and a small dose may thus cause very high local levels of radiation.

Another question is the possible long-term effects of low-level radiation. It is known on the basis of experimental radiobiology that ionizing radiation can cause genetic changes in the gametes of the adult and fetus (see Neel, 1963). Several epidemiologic studies have been performed to discover possible cancerogenic and life-shortening effects that were previously claimed to be some of the long-term effects of low-level irradiation. It seems that radiation at different times of human life, including the period *in utero,* may be partially responsible for the development of leukemia (Graham *et al.,* 1966). These very important aspects of radiation pathology fall beyond the scope of this presentation, but there are good reasons for believing that much lower doses of radiation than are needed for immediate injury may cause some of the consequences mentioned.

It is the duty of investigators in this field to meet the challenge of the atomic era and make sure that technologic development will not overwhelm human development in the future.

SUGGESTED READINGS

Review Papers

Bacq, Z. M. and P. Alexander. Fundamentals of Radiobiology, 2nd ed. Pergamon Press, London (1961).

Brent, R. L. The effect of irradiation on the mammalian fetus. Clin. Obstet. Gynecol. **3:** 928–950 (1960).

Brent, R. L. and R. P. Jensh. Intra-uterine growth retardation. *In* Advances in Teratology, (D. H. M. Woollam, ed.), Vol. 2, pp. 139–227. Logos Press, London, Academic Press, New York (1967).

Cellular Radiation Biology: Papers presented at 18th Annual Symposium on Fundamental Cancer Research. Williams & Wilkins Co., Baltimore (1965).

Hicks, S. P. Developmental malformations produced by radiation. A timetable of their development. Am. J. Roentgenol. Radium Therapy Nucl. Med. **69:** 272–293 (1953).

Hicks, S. P. and C. J. D'Amato. Effects of ionizing radiations on mammalian development. *In* Advances in Teratology, (D. H. M. Woollam, ed.), Vol. 1, pp. 195–250 Logos Press, London, Academic Press, New York (1966).

Russell, L. B. The effects of radiation on mammalian prenatal development. *In* Radiation Biology, (A. Hollaender, ed.), Vol. 1, Part II, pp. 861–918, McGraw-Hill, New York (1954).

Russell, L. B. and W. L. Russell. An analysis of the changing radiation

response of the developing mouse embryo. J. Cellular Comp. Physiol. **43**: Suppl. 1: 103–149 (1954).
Wolff, S. Radiation genetics. Ann. Rev. Genet. **1**: 221–244 (1967).

Special Articles

Dewey, W. C. and R. M. Humphrey. Radiosensitivity and recovery of radiation damage in relation to the cell cycle. *In* Cellular Radiation Biology, pp. 340–375. Williams & Wilkins Co., Baltimore (1965).

Djordjevich, B. and L. J. Tolmach. Responses of synchronous populations of HeLa cells to ultraviolet radiation at selected stages of the generation cycle. Radiat. Res. **32**: 327–346 (1967).

Evans, H. J. The response of human chromosomes to ionizing radiations: *in vitro* studies. Roy. Soc. Edinburgh Proc. Sect. B **70**: 132–151 (1968).

Gefter, M. L., A. Becker, and J. Hurwitz. The enzymatic repair of DNA, I. Formation of circular DNA. Proc. Nat. Acad. Sci. U.S. **58**: 240–247 (1967).

Goldstein, L. and D. P. Murphy. Etiology of ill-health of children born after maternal pelvic irradiation. II. Defective children born after postconception pelvic irradiation. Am. J. Roentgenol. Radium Therapy Nucl. Med. **22**: 322–331 (1929).

Graham, S., M. L. Levin, A. M. Lilienfeld, L. M. Schuman, R. Gibson, J. E. Dowd, and L. Hempelmann. Pre-conception, intra-uterine and post-natal irradiation as related to leukemia. Nat. Cancer Inst. Monograph **19**: 347–371 (1966).

Kaplan, S. J., R. Rugh, and R. K. White. The behavior of 100 day old male rats resulting from fetal X-irradiation. Atompraxis **9**: 11–16 (1963).

Kaplan, H. S., K. C. Smith, and P. A. Tomlin. Effect of halogenated pyrimidines on radiosensitivity of *E. coli.* Radiation Res. **16**: 98–113 (1962).

Miller, R. W. Delayed effects occurring within the first decade after exposure of young individuals to the Hiroshima atomic bomb. Pediatrics **18**: 1–18 (1956).

Neel, J. V. Changing perspectives on the genetic effects of radiation. Charles C Thomas, Springfield, Ill. (1963).

Plummer, G. Anomalies occurring in children exposed *in utero* to atomic bomb in Hiroshima. Pediatrics **10**: 687–693 (1952).

Puck, T. T. Cellular interpretation of aspects of the acute mammalian radiation syndrome. *In* Cytogenetics of Cells in Culture, (R. J. C. Harris, ed.), pp. 63–77. Academic Press, London (1964).

Russell, L. B. Death and chromosome damage from irradiation of preimplantation stages. *In* Ciba Foundation Symposium on Preimplantation Stages of Pregnancy, (G. E. W. Wolstenholme and M. O'Connor, eds.), pp. 217–241. J. & A. Churchill Ltd., London (1965).

Scott, D. and H. J. Evans, X-Ray-induced chromosomal aberrations in *Vicia faba:* Changes in response during the cell cycle. Mutation Res. **4**: 579–599 (1967).

Sinclair, W. K. and R. A. Morton. Variations in X-ray response during the division cycle of partially synchronized Chinese hamster cells in culture. Nature **199**: 1158–1160 (1963).

Tolmach, L. J., T. Terasima, and R. A. Phillips. X-Ray sensitivity changes during the division cycle of HeLa S3 cells and anomalous survival kinetics of developing microcolonies. *In* Cellular Radiation Biology, pp. 376–393. Williams & Wilkins Co., Baltimore (1965).

Weiss, B. and L. Tolmach. Modification of X-ray- induced killing of HeLa S3 cells by inhibitors of DNA synthesis. Biophys. J. **7**: 779–795 (1967).

Wolff, S. and H. E. Luippold, Metabolism and chromosome-break rejoining. Science **122**: 231–232 (1955).

Yamazaki, J. N., S. W. Wright, and P. M. Wright. A study of the outcome of pregnancy in women exposed to the atomic bomb blast in Nagasaki. J. Cellular Comp. Physiol. **43**: Suppl. 1, 319–328 (1954).

8

Chemical Teratogenesis

Chemical teratology as a distinct branch of this discipline dates to the 1930s, if some sporadic older reports are disregarded. Hale (1933) discovered that pigs are born eyeless if their mothers are deprived of vitamin A during pregnancy. Hale's discovery was rapidly followed by reports of congenital malformations produced experimentally as a result of dietary deficiency or gestational drug treatment.

The teratogens used in the early experiments were diverse and information emerged slowly about the mechanisms of drug-induced congenital defects. More light was thrown on the subject when drugs with known effects on cellular metabolism were used. In this respect, alkylating agents and antimetabolites proved useful.

Some years ago, those working in therapeutic medicine may have believed that the risk of drug-induced congenital defects in man was relatively small, although many drugs were teratogenic to experimental animals. Then the teratogenic activity of an apparently innocuous and very mild sedative, thalidomide, focused public attention on the potentially harmful effects of drugs on the developing embryo. The scientists working in this field had to face the questions raised by the public: Which drugs can be safely used during pregnancy? Why were new drugs not tested so that disasters like the thalidomide-induced anomalies could never occur? Despite a great amount of experimentation and other investigations, these questions cannot be answered even today, although several years have elapsed since the thalidomide incident.

The list of drugs harmful to the animal embryo is long, and if the experiments are performed under appropriate conditions and with suitably selected animals almost any drug or chemical may be teratogenic (see Fave, 1964; Cahen, 1966).

The drugs themselves and the experimental conditions, dosage, and so forth in animal experiments are such that the results cannot be directly compared to experiences in human medicine. The list of drugs established to have a teratogenic effect in man is short. In addition to thalidomide, some antitumor agents, such as aminopterin, cause congenital anomalies and some gestagenic hormones interfere with the sexual differentiation of the human fetus, causing pseudohermaphroditism (see Chapter 6). A variety of other drugs are suspected to be teratogenic, but definite proof is still lacking.

GENERAL PRINCIPLES

Drug treatment of a pregnant female may theoretically affect the embryo in one of three ways. It may improve its chances of survival and development by alleviating some disease or other harmful condition in the mother. Secondly, it may injure the fetus in such a way that, although its survival is not affected, the offspring will have congenital defects. Thirdly, the drug or chemical compound may kill the fetus. The first and second alternatives are not always mutually exclusive.

The effect of a drug on the fetus is also dose-dependent, as is easy to demonstrate experimentally with free-living embryos or tissue culture methods. With mammalian embryos, however, the situation is much more complex. When the teratogenic action of a drug is evaluated, toxicity to the mother in relation to the lethal dose and anomaly-producing level in the offspring have to be considered.

Murphy (1965) studied the relationship of drug dosage to maternal toxicity, embryonic toxicity, and anomaly-producing property, using various alkylating agents and metabolic inhibitors. Some of the results of her studies are shown in Fig. 8.1.

It is evident that the teratogenic range of different drugs in relation to lethal effect varies considerably. Some drugs, such as aminopterin in these studies, do not cause anomalies at a dosage level below fetal LD_{50}, while the teratogenic range of other drugs is wide. As seen in Fig. 8.1, some drugs, such as 2-ethylenamino thiadiazole, are toxic to the mother and hence their teratogenicity is restricted even though the anomaly-producing dosage range is relatively wide below the latter LD_{50}.

Comparable results with other chemicals, which probably have a completely different mechanism of action, have been reported by Beck and Lloyd (1965). These workers, using diazo dyes, were able

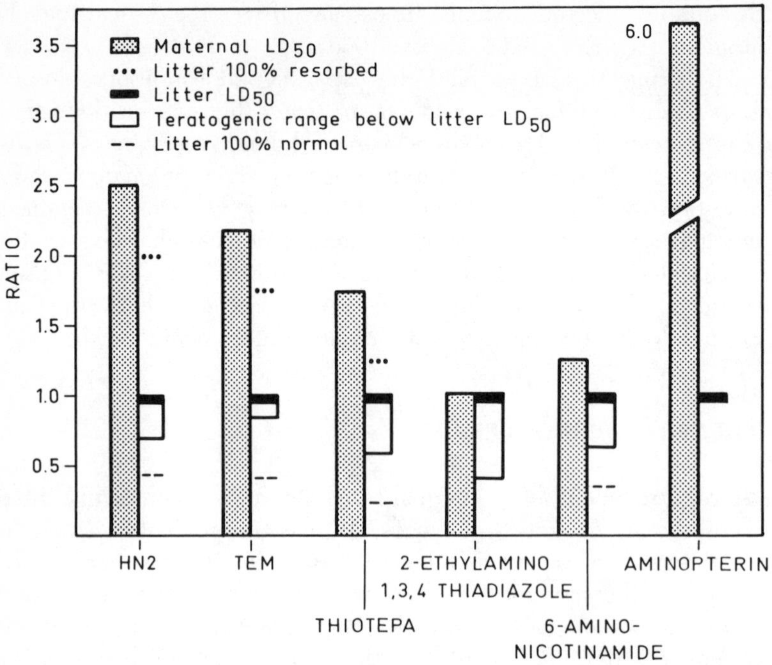

Fig. 8.1 Ratio of dosage of alkylating agents and metabolic inhibitors given to pregnant rats on day 12 of gestation to produce various effects on the mother and her fetus to estimated litter LD_{50}. (After L. Murphy. *In* Teratology, Principles and Techniques. J. G. Wilson and J. Warkany [eds.], p. 145. The University of Chicago Press, Chicago and London [1965].)

to show that the dosage range between the anomaly-producing and embryo-killing effects varied considerably with different diazo derivatives (Fig. 8.2). Trypan blue causes anomalies in about 20 percent of the fetuses at the most effective dosage whereas Niagara blue does not cause anomalies in any appreciable proportion, but when the dosage is increased the fetuses will die.

From these observations it can be concluded that a dangerous teratogen is one which is relatively atoxic to the mother and which has a wide teratogenic dosage range, that is, produces a high percentage of anomalies at dosage levels which are not lethal to the embryos.

The Route of Teratogenic Action

It is generally thought that most chemicals administered to the pregnant animal or avian embryo exert their harmful action directly on the embryo. Obviously, this is what happens if the drug is given to free-living amphibian and fish embryos or chick embryo explants. When dealing with mammals, some other factors have to be taken into account. Maternal metabolism may modify the drug before it reaches the fetus. The mammalian placenta restricts the free passage of certain large molecular substances and has an active metabolism of its own, which may alter the structure of various substances, such as steroid hormones. Both maternal metabolism and the placenta supposedly protect the developing fetus, but if they are the primary targets of a drug their impaired function may be harmful. Some examples of such indirect action are given below.

Trypan blue and some related dyes cause severe defects in fetuses of various animals and have been the object of intensive studies about the several aspects of chemical teratology. When trypan blue is injected into pregnant rats or some other animals, it accumulates in the maternal tissues and in the epithelium of the yolk sac placenta,

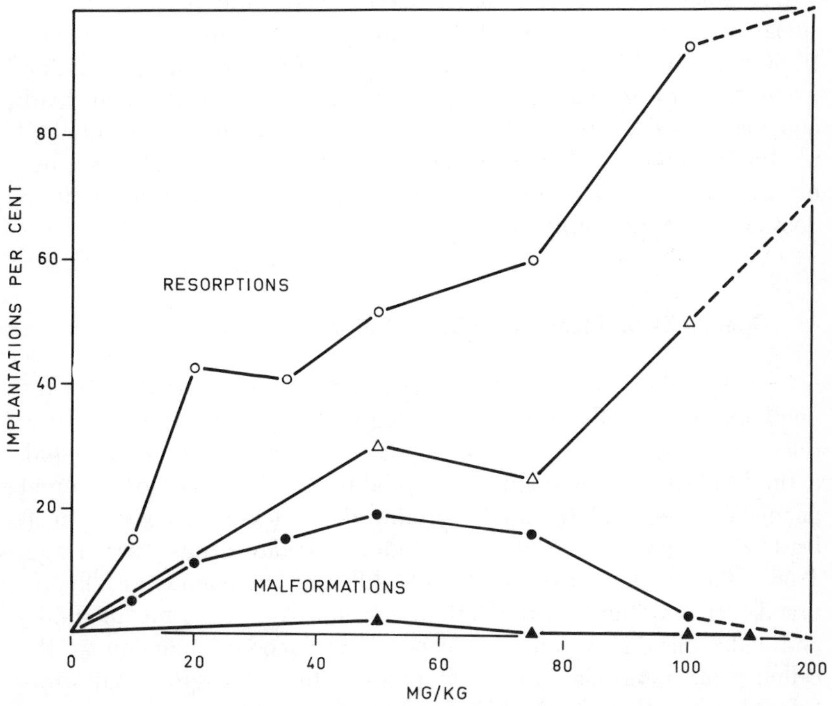

Fig. 8.2 The effect of trypan blue on fetal mortality (○) and its teratogenicity (●) as compared to the same effects of Niagara blue 4 B (triangles) at different dosage levels in Wistar rats. (After P. Beck and J. Lloyd. **In** Embryopathic Activity of Drugs. J. M. Robson, F. M. Sullivan, and R. L. Smith [eds.], Churchill, London [1965] p. 13.)

but not in any other fetal membranes or tissues. Thus, it seems that the teratogenic action is an indirect one, and theories based on altered maternal serum proteins or metabolism have been presented to account for the teratogenicity of the dye (see Beck and Lloyd, 1966). However, Beck and his collaborators (1967) have presented a more plausible theory to explain the mechanism of the teratogenic action of trypan blue. The teratogenic action of the dyes is restricted to the first 8 to 10 days of rat pregnancy. After this, no anomalies are induced. This time coincides with the establishment of the final chorio-allantoic placenta in the rat. Prior to this, the implanted embryo obtains its nourishment through the visceral yolk sac epithelium. The dye concentrates on the yolk sac epithelium and inhibits lysosomal enzymes in the epithelium. It is assumed that the lysosomal enzymes perform an important function in transferring nutrients from the yolk to the embryo and thus the teratogenic action of azo dyes consists of impairment of embryonic nutrition. When yolk sac function ceases at the time of placental circulation, the dyes are no longer teratogenic.

Another example of indirect action, where the placenta is presumably the primary target of the teratogen, is the effect of 5-hydroxytryptamine on the embryo (Robson *et al.*, 1965). This biologic amine, when given to pregnant mice as a single injection, causes fetal death and malformed offspring. The effect has been attributed to impaired placental blood circulation, which causes fetal anoxia and deficiency of other nutrients. The 5-hydroxytryptamine concentration in the fetus does not rise appreciably.

Specificity of Teratogenic Drug Action

One of the most perplexing aspects of teratology is that, on the one hand, chemically and pharmacologically different drugs may produce the same structural defect in the fetus and, on the other hand, a single chemical compound may produce a wide array of different anomalies. Cleft palate can be produced in mouse embryos with at least 20 different treatments, including various drugs (see Dagg, 1966). Among these are such widely different compounds as 5-fluorouracil, cortisone and its derivatives, vitamin A, 6-aminonicotinamide, and galactoflavin, which all have different modes of action at the cellular and molecular levels yet produce the same gross malformation. On the other hand, drugs such as trypan blue and large doses of vitamin A, when appropriately administered, produce maldevelopment in almost all organ systems of the fetus. At present, a common denominator for these apparent illogicalities might be sought in the molecular basis of development. Theoretical considerations of some

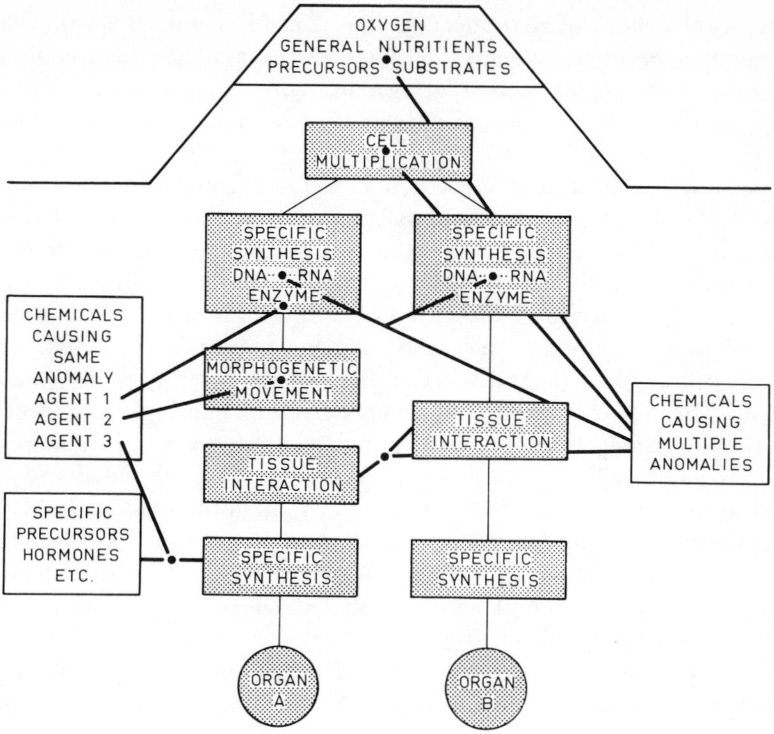

Fig. 8.3 Hypothetic possibilities of the primary targets of different teratogenic drug actions.

of the ways in which a single teratogen may cause multiple defects or several agents induce the same anomaly are presented in Fig. 8.3.

It is reasonable to assume that the developing organism presents a multitude of metabolic targets vulnerable to attack by chemicals. The resulting array of morphologic abnormalities depends on whether the chemical acts on a pathway common to many developmental events or at a specific site involved in the course of differentiation of a single organ. This reasoning implies a specificity between the teratogenic agent and its receptor site in the organism, but because the same receptor is shared by many organs while a single organ has several receptor sites, the resulting anomalies do not always reflect the specificity of the teratogen.

The Effect of Several Drugs Administered Simultaneously

In addition to the proper time of administration and dosage, a number of other, apparently unrelated, factors may modify the teratogenic effect of a drug. This interplay of factors is possible at the genetic or environmental level (see the latter part of this chapter), but the simultaneous action of two or more environmental factors may also have an unexpected effect. It has been suggested that when several factors, for instance chemicals, act on the developing organism at one time, the effect is one of the following (Runner and Dagg, 1960): nil effect, additive effect, potentiating effect, or interfering effect. A nil effect is observed if the simultaneous use of two teratogens gives the same result as would have been obtained with the more efficient drug alone. An additive effect means that the number of malformations is the same as their sum would have been if the two drugs had been used separately. If the teratogens potentiate each other's effect, their simultaneous use results in a higher incidence of malformations than would be expected from mere summation of their separate effects. Interference means that the combined use of teratogens leads to a lower incidence of malformations than with the use of either substance alone. Some examples of these possibilities, as revealed by animal experiments, are given on the following pages.

The simplest example of a combined effect of two teratogenic treatments is obtained by administering cortisone in combination with restricted calory uptake at the critical period of development in a susceptible mouse strain. The results of such an experiment are presented in Table 8.1.

TABLE **8.1**
The Effect of Cortisone and Restricted Diet, Separately and Combined, on the Incidence of Cleft Palate in A/J Mice.[a]

Treatment	Cleft Palate Percentage
Restricted diet	5.6
½ mg cortisone	1.4
½ mg cortisone and restricted diet	36.8
1 mg cortisone	12.0
1 mg cortisone and restricted diet	50.7

[a] After H. Kalter: in Teratology, Principles and Techniques, (J. G. Wilson and J. Warkany, eds.), University of Chicago Press (1965), p. 57.

In the above experiment the two treatments potentiate each other because the resulting incidence of cleft palate exceeds the mere sum of the effects of each component alone.

Examples of the additive and nil effects derived from the studies of Runner and Dagg (1960) are given in Table 8.2.

TABLE **8.2**
Effect of Fasting and Various Drugs on the Frequency of Vertebral Anomalies in 129/RrJ Mice, when Applied on Days 8 and 9 of Pregnancy.[a]

Treatment	Food ad libitum Percentage Malformed	Fasting Percentage Malformed	Type of Effect
None	0	24	
Trypan blue	16.9	43.5	Additive
9-Methyl folic acid	13.3	18.5	Nil
Iodoacetate	62.6	66.1	Nil

[a] After M. N. Runner and C. P. Dagg: National Inst. Monograph No. 2, 1960, p. 41.

Of great theoretic interest and possible practical importance in human teratology are environmental factors which cause malformations when used at subthreshold levels of the individual components. There are some observations that a treatment, which in itself is not teratogenic, may intensify the effect of a teratogen. Even more important are teratogens which cause malformations only in combined use. An example of this has been reported by Landauer and Clark (1964). If chick embryos are treated with low doses of sulphanilamide or 6-aminonicotinamide, no malformations result. In higher doses both compounds are teratogenic, however. In Table 8.3 the effect of combined treatment with these drugs is presented.

TABLE **8.3**
The Effect of Sulphanilamide and 6-aminonicotinamide on Chick Embryos when Injected at 96 Hours of Incubation Singly or in Combination.[a]

Sulphanilamide μg/egg	300	300	
6-aminonicotinamide μg/egg		1.5	1.5
Number treated	106	108	105
Survivors of day 13	101	97	99
Normal percentage	99.0	19.6	100
Micromelia percentage	0	80.4	0
Beak malformations percentage	0	55.7	0
Other malformations	1.0	0	0

[a] After W. Landauer and E. M. Clark: Nature **203:** 527, 1964.

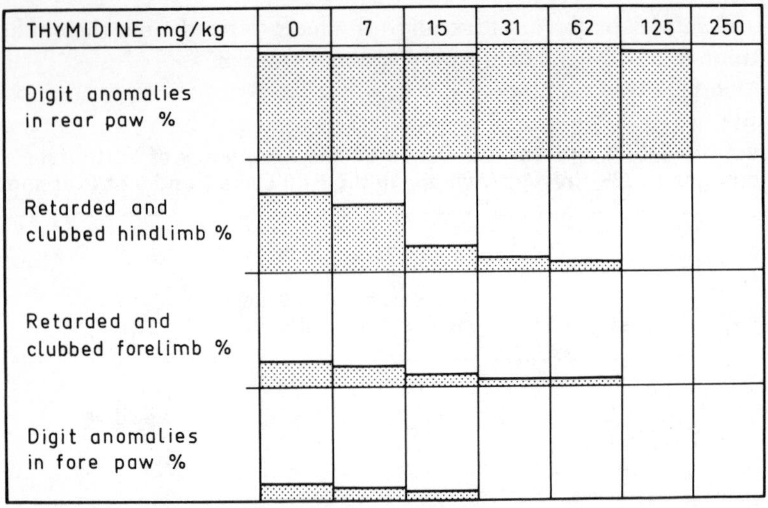

Fig. 8.4 Protective effect of different amounts of thymidine against a treatment of 500 mg/kg of 5-chlorodeoxyuridine, when injected simultaneously into 12-day pregnant rats. Shaded area shows the percentage of malformations. (Based on data from S. Chaube and M. L. Murphy. Cancer Res. **24:** 1986 [1964].)

In some instances the embryo can be protected against the teratogenic action of a drug. The effect of vitamin antagonists and some metabolic inhibitors can be counteracted by simultaneous use of the vitamin or specific substrate. An example of such an effect is given In Fig. 8.4. In this case a physiologic competitor of the nucleic acid precursor analog 5-deoxycytidine can completely counteract the teratogenic effect if given in sufficiently high doses.

The case is more complicated if two teratogens prevent or lessen each other's activity—an interference effect in the true sense. Examples of such effects are given by Landauer and his associates (Landauer and Clark, 1962, 1964). Two nicotinic acid amide analogs, 6-aminonicotinamide (6-AN) and 3-acetylpyridine (3-AP), are both teratogenic to chick embryos and their effect on development is expressed in different types of anomalies. 6-AN produces dwarfing micromelia, and parrot beak, whereas 3-AP mainly produces muscular hypoplasia. It was found that, when administered in the appropriate ratio, 3-AP almost completely abolished the defects typically produced by 6-AN. On the other hand, the defects due to the action of 3-AP tended to be intensified in the presence of 6-AN. Similar, but more complicated results were obtained if treatments with 3-AP and sulphanilamide were combined. 3-AP prevented the teratogenic action of sulphanilamide, that is, formation of micromelia and parrot beak. The effect of sulphanilamide on the teratogenic action of 3-AP was more complicated. If the 3-AP treatment at 72 hours of incubation was preceded by sulphanilamide application 24 hours earlier, the effect of 3-AP on muscular development was reduced. If the experiment was repeated at 96 hours of incubation, the preceding or simultaneous sulphanilamide treatment exaggerated the effect of 3-AP. These experiments show the complexity of the interaction of drugs when their targets are the constantly changing receptor sites of the developing organism.

Effect of Genotype on Teratogenic Response to Chemicals

Different animal species respond to the embryopathic action of drugs in different ways, even if exposed to their action in comparable developmental stages and drug dosages. A widely employed teratogen, cortisone, produces cleft palate and other malformations in the mouse and the rabbit, but the rat is remarkably resistant to this compound. Similarly, thalidomide is highly teratogenic in man and in the monkey, but less so in the mouse and the rabbit, and it has not been reported to produce typical limb abnormalities in the rat. With some drugs, however, the teratogenicity seems to be the same over a wide array of animal species. Hypervitaminosis-A is equally teratogenic in the rat, mouse, rabbit, golden hamster, and guinea pig.

Within the same species a drug may show differential teratogenicity in different strains. This has been demonstrated in numerous experiments on inbred strains of mice. In the experiments of Fraser and Kalter and others, the susceptibility of different inbred strains of mice to cleft palate as a result of the action of cortisone has been beautifully worked out (see Chapter 2, Table 2.1).

The recognized fact that a given teratogenic treatment causes the same anomaly in different proportions of embryos, depending on the strain, led investigators to study the interplay of genetic and environmental factors, using crosses and backcrosses between strains differing in susceptibility. This interplay has been dealt with at some length in Chapter 3 and a couple of examples illustrating the principles of drug teratology will be cited here.

It soon became evident that the result of treatment with a teratogen depends not only on the genotype of the fetus, but also on that of the mother and the father. Goldstein (1963) studied the effect of 6-aminonicotinamide on two types of anomalies in two inbred strains of mice and crosses between them (Table 8.4).

TABLE **8.4**
The Incidence of Vertebral Fusions and Cleft Palate in Offspring of Reciprocal Crosses between 2 Inbred Strains of Mice Treated with 6-Aminonicotinamide.[a]

	Cross ♀	♂	Percentage of Vertebral Anomalies	Percentage of Cleft Palate
1	A	A	89	76
2	C57	A	67	4
3	A	C57	45	36
4	C57	C57	56	11

[a] After M. Goldstein, F. Pinsky, and F. C. Fraser: Genet. Res. **4:** 258, 1963.

Table 8.4 indicates that in at least one of the combinations the interstrain cross displays sensitivity intermediate between the pure strains (cross 2 for vertebral anomalies and cross 3 for cleft palate). On the other hand, it is evident that in addition to the genotype of the fetus, which is the same in crosses 2 and 3, maternal and paternal factors influence the outcome of the teratogenic treatment. Fetuses developing in strain A mothers are more resistant to the teratogenic effect on vertebrae than fetuses developing in C57 mothers. In regard to cleft palate, the situation is reversed. Several hypotheses can be postulated to explain this:

1. A factor for resistance transmitted in the sperm but not in the egg (cross 3 for vertebral anomalies).
2. A maternal uterine factor for resistance (cross 3 for vertebral anomalies).
3. Differences in development rates of the hybrid fetuses depending on the maternal organism, so that the fetuses of reciprocal crosses are exposed to the teratogen at different stages of development.
4. Factors for resistance (or susceptibility) handed on by the egg cytoplasm and not present in the sperm.

Experiments using reciprocal crosses and backcrosses between susceptible and resistant strains of mice combined with teratogenic treatment have shed light on the interplay between genotype and environment (see Chapter 3). The results of these experiments may be summarized as follows:

1. Susceptibility to a given teratogen seems to be controlled by several genes—maternal, paternal and fetal genotypes each playing a distinct role in response to the action of an environmental factor.
2. Genes responsible for differential susceptibility to a teratogen are organ-specific rather than systemic. In other words, genes do not lower or increase the general susceptibility of the fetus to the action of a teratogen, but work at the level of particular tissues (see Table 8.4).
3. Susceptibility to a given anomaly also seems to be controlled by several hereditary factors, which may be differentially affected by various teratogens.

A case in point is presented in Table 8.5, which shows that strain susceptibility to one defect may turn to resistance if a different teratogen is used.

The remarkable results obtained from experiments with inbred mice strains may help to explain the inconsistent results obtained when other genetically heterogeneous animals are used or the variable observations relating to man.

TABLE **8.5**

Effect of Galactoflavin and Cortisone on the Incidence of Cleft Palate in Different Strains of Mice.[a]

Strain	Percentage of Cleft Palate Produced by	
	Galactoflavin	Cortisone
A	3	100
C57	0	19
DBA	61	92

[a] After H. Kalter: in Teratology, Principles and Techniques. (J. G. Wilson and J. Warkany, eds.), University of Chicago Press, Chicago and London (1965), p. 57.

Mechanism of Teratogenic Action

The final formation of every organ is the result of a long chain of differentiative steps. These can be characterized at both the morphologic and metabolic levels of differentiation. Experimental embryology combined with molecular biology has given us new insight into the problems of differentiation by revealing the metabolic processes underlying morphologic differentiation. It is reasonable to assume that most of the chemicals that distort normal development exert their primary effect at the level of metabolic pathways, and this action is secondarily reflected at the cellular and tissue levels. The metabolic needs of an organ vary during differentiation and so does the ability of the fetus to metabolize and detoxicate drugs and other chemicals. As a consequence, even if we know the effect of a chemical on the metabolism of the adult organism, we do not always know the effect it exerts on the metabolism of embryonic tissues. Thus it is understandable that in most cases, where the effect of a chemical on the embryo is malformation or developmental arrest, the underlying metabolic disturbance is not known.

The traditional way to ascertain the mechanism of the action of teratogenic drugs is to examine the fetus histologically at different times after exposure to the drug. With this old and widely used method, it has been possible to unravel some of the pathogenetic mechanisms of drug-induced anomalies. Töndury (1964), using this method, was able to demonstrate the primary target cells in the anophthalmia induced in newt larvae by triethylene-melamine (TEM). The effect of TEM can be detected histologically about 60 hours after its administration. The result is selective cell necrosis in retinal cells at the beginning of differentiation and prevention of mitosis. As shown in Fig. 8.5 different parts of the eye are affected, depending on the stage of development. Consequently, the outcome of teratogenic treatment is different. If TEM is administered at the time when mitotic activity is taking place all over the outer layer of the neural retina, the effect is so widespread that the further development of the whole eye is prevented and total anophthalmia or a rudimentary eye results. If, on the other hand, the drug is given later at the stage when proliferation is restricted to the edges of the eye cup, destruction is limited to this area and the final injury is less marked.

Many other drugs, which do not ordinarily cause cell destruction in the adult organism, manifest their primary effect in the embryo as cell necrosis. It has been shown, for example, that vitamin A and thalidomide—which are atoxic to the mother—cause cell death in embryonic tissues.

Histologic analysis of the effect of drug-induced anomalies has yielded other significant findings, besides revealing the primary site of cell destruction. It has been shown that the primary target of a chemical need not be the site at which anomaly is later found. A single dose of vitamin A given to pregnant hamsters on the day 8 of gestation causes anencephaly, spina bifida, and ocular malformations, among other defects. The central nervous system anomalies

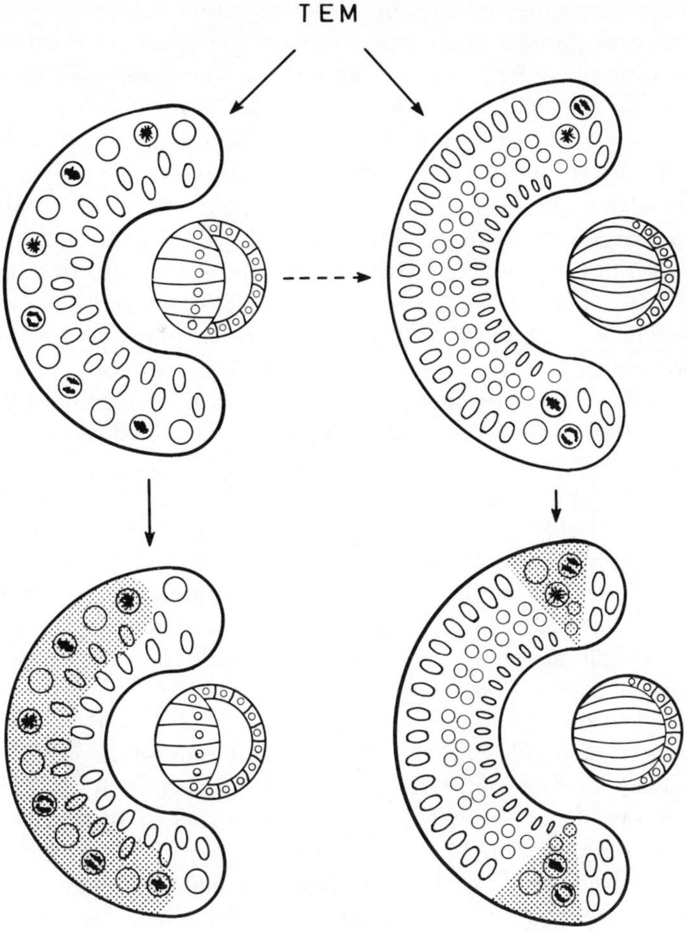

Fig. 8.5 A scheme of the effect of TEM on the newt eye. In the younger eye on the left in the figure the drug inhibits mitosis and causes cell destruction in a wide area at the outer rim of the neural retina. In older eyes the effect is restricted to the area where cell proliferation is still occurring. (Based on observations by G. Töndury. *In* Medikamentöse Pathogenese fetaler Missbildungen. Th. Koller and H. Erb [eds.], p. 4. Karger, Basel and New York [1964].)

can all be traced back to failure of the neural tube to close. Histologic analysis 12 and 24 hours after injection of the vitamin shows that the primary target is not the nervous tissue, but the cephalic somites and notochord (Marin-Padilla and Ferm, 1965). Extensive cell destruction was seen in these tissues. The failure of the neural tube to close and the nervous tissue degeneration were secondary phenomena. The somite destruction in this case probably affects the development of the structures which normally support and protect the central nervous system, that is, the vertebrae and some of the skull bones. Thus the final anomaly, anencephaly, is the result of a chain of developmental errors and the primary disturbance can be traced back to somite destruction. Another, beautifully analyzed example of a similar mechanism is described on page 93. In this case, the destruction of cells at the critical time and critical site leads to limb anomalies. In this example the normal timing of the inductive process is disturbed and although the destroyed cells are replaced by regeneration, asynchrony of the subsequent tissue interaction results in malformation.

In several instances the immediate effect of a teratogenic chemical is not so clearly visible as in the examples mentioned above. If the fetus is analyzed histologically, only indirect changes are seen. There may be retardation of morphogenetic movements, as seen in the closure of the palatal shelves, or improper development of organ primordia without any apparent cause, or there may be total failure to form a primordium at the right time. The analysis of abnormal development must then focus on the metabolism of the differentiating tissues and on any specific metabolic errors that may result from the teratogenic influence.

In view of the primary importance of nucleic acid metabolism in development and differentiation, it is of interest to analyze the teratogenic effects of substances known to interfere with nucleic acid synthesis and function. In Fig. 8.6 the action sites of several of these compounds are presented.

These compounds have found their way into teratology by virtue of their antiproliferative action and possible value in cancer chemotherapy. Several alkylating agents and other antimetabolites have proved highly teratogenic when given to pregnant laboratory animals (see Fig. 8.1). Of all these compounds, Actinomycin-D(AM) is the one in which the relationship of function at the molecular level to the effect on morphologic differentiation has been most thoroughly studied. Actinomycin prevents DNA-dependent RNA synthesis by virtue of its capacity to bind the guanyl residues of DNA. Very small doses of this antibiotic (75μg/kg), when given to the pregnant rat at 7 to 9 days of gestation, produce severe malformations in virtually all organ systems of the embryo, suggesting some general mechanism of action important to most developing tissues (Tuchmann-Duplessis and Mercier-Parot, 1960). The effect of actinomycin on differentiation has been tested in numerous isolated developmental systems; a single example will serve here—the formation of nephric tubules in the mouse kidney explant (see Saxén *et al.,* 1968). In an

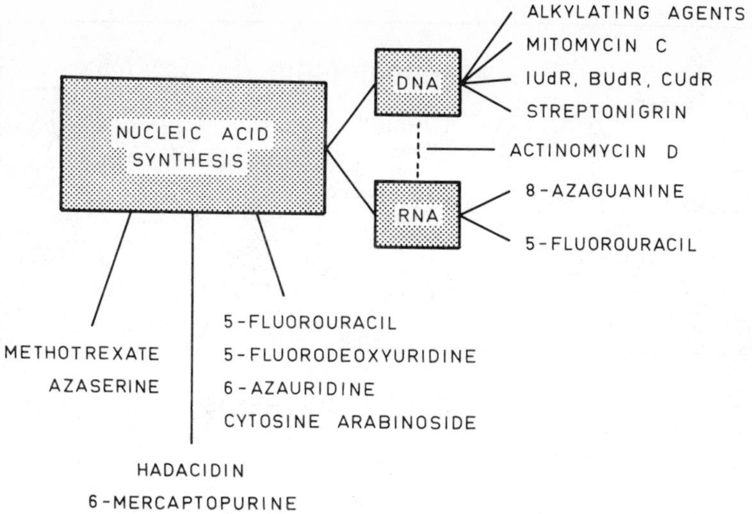

Fig. 8.6 Schematic presentation of the action sites of different nucleic acid antagonists. (Modified from D. A. Karnofsky. *In* Teratology, Principles and Techniques, J. G. Wilson and J. Warkany, eds. p. 185. University of Chicago Press, Chicago and London [1965].)

organ culture of nephrogenic mesenchyme, tubule formation is initiated by introducing a piece of spinal cord as inductor (Fig. 2.14). When the explant is exposed to actinomycin in such low concentrations (0.05 μg/ml) that it does not kill the cells, the antibiotic totally prevents tubule formation, providing that it is given not less than 15 hours before the first indication of tubule differentiation would be anticipated. Later application can no longer inhibit tubule differentiation. When this result is compared with data of RNA synthesis, it is seen that active tubule differentiation is preceded by an increase in RNA synthesis, as measured by uridine uptake (Fig. 8.7).

AM suppresses uridine uptake, its effect reaching a maximum at the period of maximal RNA synthesis in normal development. The results can be interpreted as follows: Kidney mesenchyme gets ready for tubule morphogenesis by synthesizing RNA in large quantities. Actinomycin inhibits synthesis of RNA, particularly ribosomal RNA, and, consequently, tubule differentiation. If the RNA synthesis needed for tubule formation is already completed, AM is no longer capable of preventing tubule morphogenesis. It has been suggested that the AM effect might be a reflection of the inhibition of gene-activated differentiation.

Interference with other types of macromolecular synthesis has also been suggested as the primary target of teratogenic agents, and one possible mechanism is briefly mentioned here. Several substances

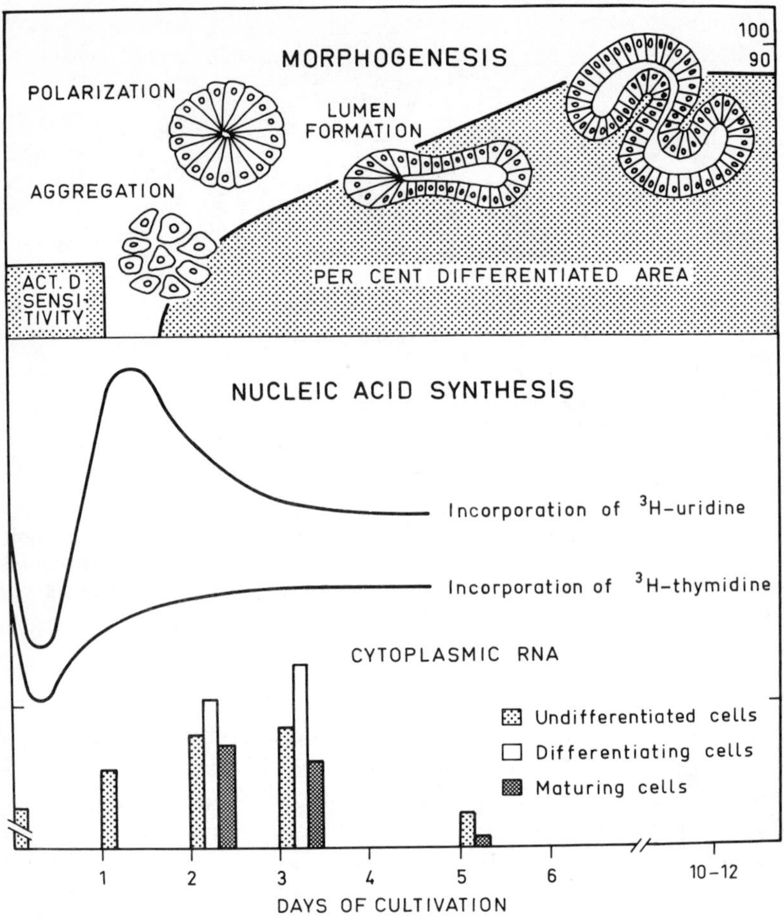

Fig. 8.7 Schematic drawing of morphogenesis and RNA metabolism in a mouse kidney explant and the actinomycin-sensitive period of morphogenetic differentiation of the tubules. (After L. Saxén *et al.* Advances in Morphogenesis vol. 7, p. 251, Academic Press, New York and London [1968].)

cause cleft palate in the fetus. Three of them, cortisone, salicylamide, and vitamin A, have as a common denominator the prevention of synthesis or accumulation of acid mucopolysaccharides in tissues under various experimental conditions. Both cortisone and salicylamide have been shown to decrease the uptake of radioactive sulfur in the sulfated mucopolysaccharides in the palatal shelves of mouse embryos (Larsson and Boström, 1965). This result, however, could not be confirmed in rat embryos treated with vitamin A, although this substance caused cleft palate in a high proportion (Kochar and Johnson, 1965).

At present, conclusive evidence of the teratogenic mechanisms of action of most of the drugs used experimentally is still wanting.

THALIDOMIDE EMBRYOPATHY

In the preceding paragraphs some outline of the principles of drug-induced teratology, drawn from experiments with laboratory animals, are given. The potential teratogenic danger of drug therapy in man was realized a number of years ago. Actual evidence of drug-induced malformations in human beings was scanty, however. The drug used in animal experiments are in general toxic, and many of them have very limited clinical use, if any. Moreover, very large doses have to be used to produce malformations in animals. Therefore, it was a shock to the medical world in 1961 when both Lenz in Germany and McBride in Australia reported a possible connection between the use of a mild sedative, thalidomide (Fig. 8.8) and an increase in certain previously rare but severe malformations in newborn babies. This discovery was soon confirmed in Britain and other West European countries, as well as in other parts of the world where thalidomide had been on sale.

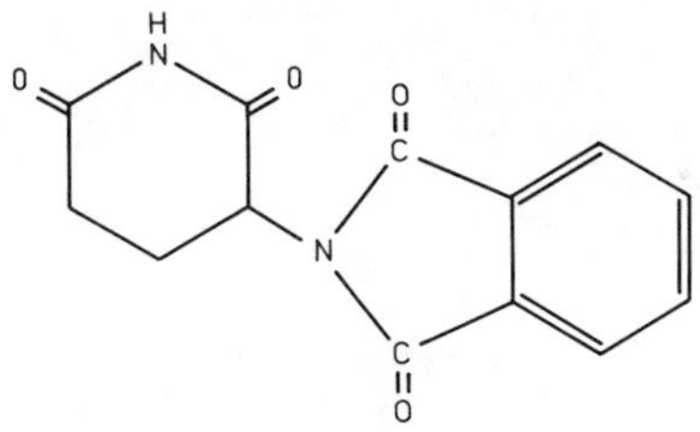

Fig. 8.8 Structural formula of thalidomide.

The most conspicuous changes in the thalidomide syndrome were various limb defects, ranging from total absence of limbs—amelia—to minor defects in the fingers or toes. A common and typical anomaly was phocomely (Fig. 8.9), which means that a more or less anomalous hand or foot, resembling the flipper of a seal, is connected with the body directly or by a short abnormal bone. Such limb anomalies were accompanied in various proportions by other defects of the ear, heart or intestinal organs.

There was strong evidence incriminating thalidomide as the factor responsible for the appearance of the typical malformation syndrome. In Fig. 8.10 are presented data of the incidence of typically

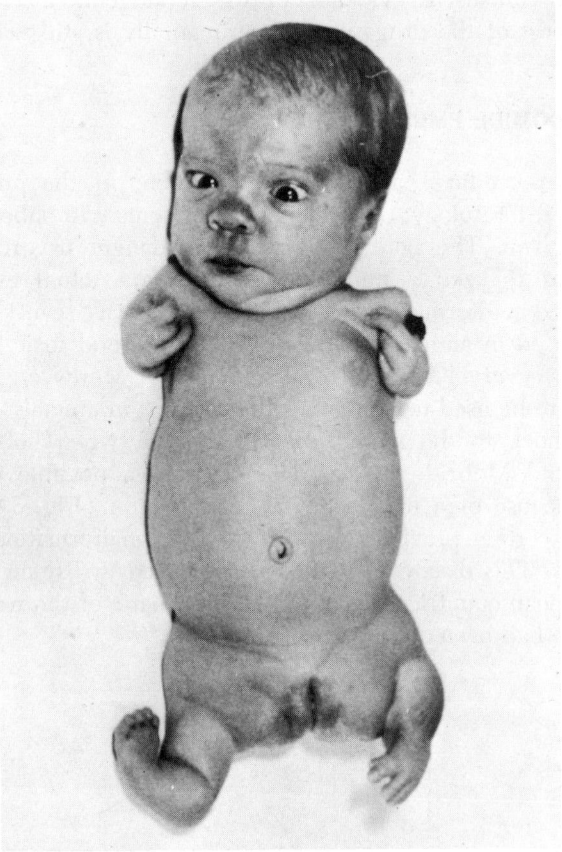

Fig. 8.9 Typically malformed newborn baby with phocomelia in all limbs. (From I. Väänänen, and T. Joki. Ann. Paediat. Fenn. **9:** 65 [1963].)

malformed children in an area in both England and Japan, together with the amount of thalidomide on the market. It is evident that the thalidomide wholesale figures parallel the incidence of these malformations in both countries and that just as the peak supply occurs later in Japan than in England, so does the peak incidence of malformations.

In countries where thalidomide was not marketed, the most significant being the United States, no increase of such anomalies was reported. It should be strongly emphasized that the discovery of a correlation between thalidomide and malformations was largely due to the severity of the anomalies caused by the drug and their previous rarity. It is also very likely that the teratogenicity of thalidomide was very pronounced, that is, the risk of the birth of a deformed child was high if thalidomide was taken during the critical period of pregnancy.

The thalidomide disaster changed the attitude of the medical profession toward drug-induced malformations. Whereas previously all tests had been performed with laboratory animals and the results used as a basis for predicting what would happen if the drug were consumed by pregnant women, the case with thalidomide was reversed. The course of events in the thalidomide disaster also enabled

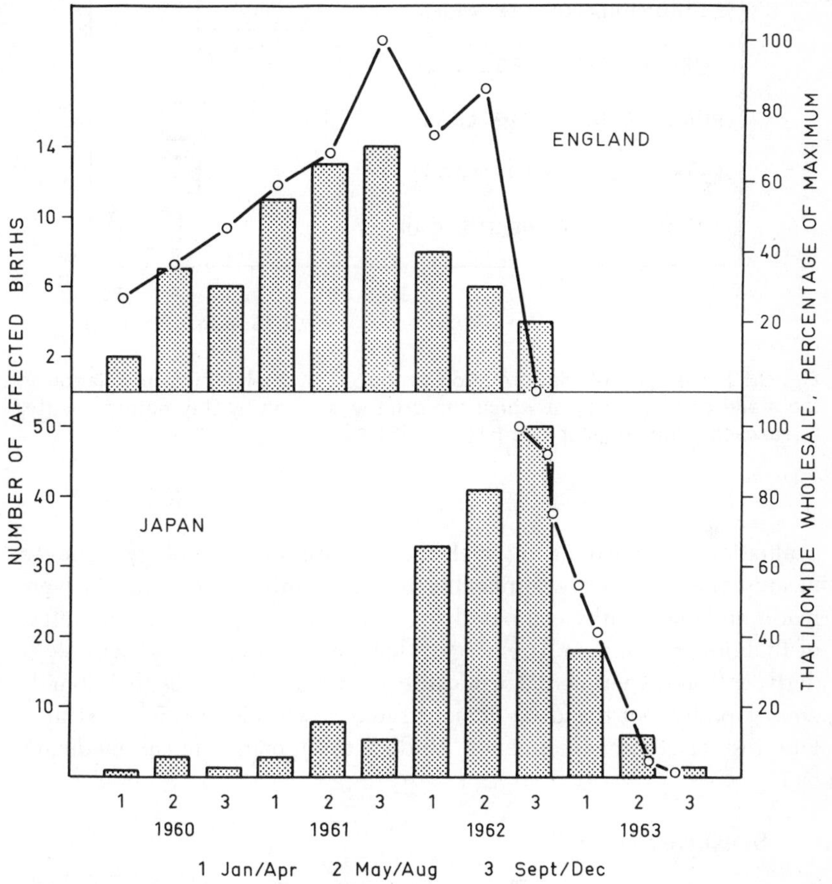

Fig. 8.10 Frequency of affected births in England (upper figure) and in Japan (lower figure) compared with the supply of thalidomide nine months (England) and eight months (Japan) earlier. (After R. W. Smithels and Leck, I. Lancet **i:** 1095 [1963] and Kajii, T. Ann. Paediat. **205:** 341 [1965].)

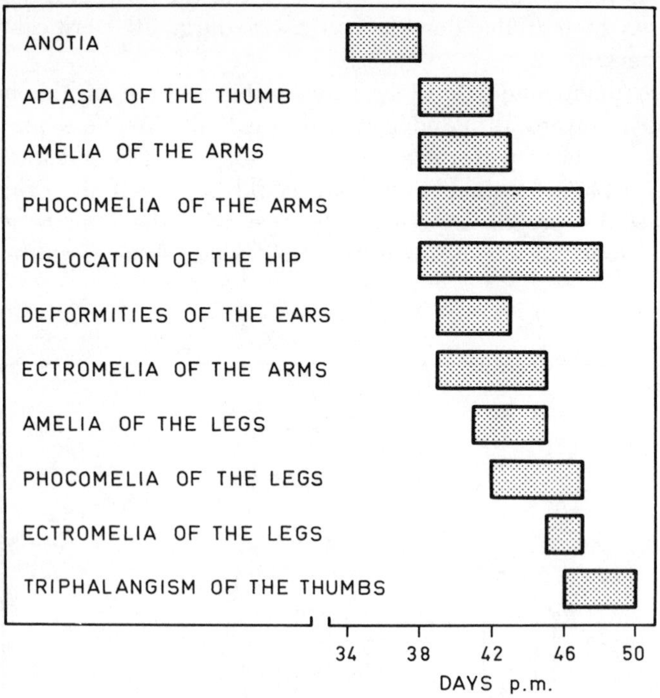

Fig. 8.11 The type of malformation produced by thalidomide in relation to the stage of pregnancy at which the drug was taken by the mothers. (After E. Nowack. Humangenetik **1:** 516 [1965].)

evaluation of the extent to which the principles of teratology achieved in animal studies were applicable to man. Animal experiments were made and the results compared to what was known about the effect of thalidomide on man. It emerged that some of the laws of teratology derived from experimental work also held true of human thalidomide embryopathy, but this drug demonstrated how difficult it is to extrapolate results from experimental studies to human clinical medicine.

Sensitive Period

Nowack (1965) reviewed the records relating to a large number of mothers who had given birth to a typically malformed child. He selected a series in which the time of thalidomide consumption was well documented and correlated this to the type of anomaly in the child. The results of his study are presented in Fig. 8.11. It is evident that the embryopathic activity of this drug tallies very accurately with the scheme of short sensitive periods of different organs during organogenesis.

Pathogenesis

The gap between the dose causing a harmful effect in the mother and the embryopathic dose is extremely wide. It has been reported that large overdoses of thalidomide are tolerated without any serious side-effects. On the other hand, the amounts taken by pregnant mothers were usually small and fell within the therapeutic range of 100 to 300 mg as a single dose. No harmful effects on the mothers who took the drug during pregnancy have been reported, but it is known that thalidomide may cause neuropathic symptoms after prolonged use.

Almost all the aspects of chemical teratology discussed in this chapter have been tested with thalidomide in animals, but the results obtained were somewhat unexpected in the light of the known effects of this drug in man. The first animals used for teratologic experiments were mice and rats. It turned out in several studies that thalidomide did not cause anomalies in any appreciable amount in rat embryos and the reports agree that it does not produce leg anomalies in the rat. Large quantities of the drug (500–1000 mg/kg) caused fetal death and in a number of experiments anomalies were found in the vertebrae and the sternum. The results with mice were similar, although mouse embryos seemed to be slightly more sensitive to the embryotoxic and embryopathic effects of thalidomide. Typical leg abnormalities were not found in most of the studies. Skeletal malformations in mice were found more often than in rats and the same was true of visceral and nervous malformations.

The rabbit was the first test animal found to show leg anomalies comparable to those occurring in man. Shortening of the limbs due to reduction in the length of the long bones and hypoplasia of the radius and ulna occurred in a high percentage of cases, as well as corresponding defects in the fibula and tibia. Anomalies in the paws were also frequent. These findings have been confirmed by several workers independently. Phocomely, a characteristic anomaly in the human syndrome, was not found in rabbits. There are incidental reports on other rodents and laboratory animals, but the results have been too contradictory to allow any conclusions to be drawn. Some results with monkeys merit attention. In small series of *Cynomolgus* monkeys (Delahunt and Lassen, 1964) and baboons (Hendrickx *et al.*, 1966), anomalies similar to those occurring in man were reported, including amelia and phocomelia. In rhesus monkeys thalidomide causes sterility and abortions by killing the embryos in early stages of development (Lucey and Behrman, 1963).

Thalidomide has also been tested on avian embryos as well as on some lower animals, where its effect is not directly related to the problems of mammalian teratology. The chick embryo is interesting, because thalidomide causes much the same anomalies in the chick as in the rabbit and man. On the other hand, anomalies frequently occur spontaneously in chicks and inert insoluble particles such as sand and talc, when injected into the egg yolk, cause anomalies similar to those produced by thalidomide, which make the results difficult to interpret.

The possibility of a synergistic action of thalidomide with other teratogens has also been tested. One interesting result which may give some clue to the mechanism underlying thalidomide embryopathy deserves mention. Fratta *et al.* (1965) studied the effect on rat embryos of thalidomide combined with deficiency of several vitamins. Of a number of vitamin deficiences tested, lack of panthothenic acid was the only one which intensified the effect of thalidomide on rat fetuses. This combination also caused limb anomalies, which thalidomide alone does not produce in the rat. It may be significant that thalidomide has been claimed to lower the panthothenic acid content of the chicken liver and that injury of the rabbit liver with carbon tetrachloride enhances the embryopathic activity of thalidomide (Heine *et al.*, 1964; Toivanen *et al.*, 1964).

The metabolism of thalidomide in the human and animal organisms has been studied extensively. Thalidomide is an unstable compound in neutral or slightly alkaline solution. Under these conditions it is readily hydrolyzed, one of the imide bonds (..NH.O..) breaking with resultant formation of carboxylic acids. When these products of spontaneous hydrolysis are compared to the metabolites excreted in human and animal urine, they are found to be qualitatively and quantitatively about the same. When tested on rabbits, none of the hydrolysis products are embryopathic (Keberle *et al.*, 1965). These findings invalidate many of the earlier experimental studies with thalidomide because insoluble thalidomide was solubilized on several occasions in alkaline solution.

In chemical structure, thalidomide resembles glutamine and glutamic acid, which are known to be precursors of various biologically important substances. It has been suggested that thalidomide might serve as an analog for glutamine or other group B vitamins, but no evidence has been found to prove that it acts as a vitamin antagonist. Thalidomide has also been tested in regard to its ability to inhibit cell growth in tissue culture, interfere with amphibian tadpole metamorphosis, interfere with immunologic phenomena, and so forth. So far none of these experiments have disclosed the teratogenic mechanism of this drug.

Conclusions

It may be concluded that thalidomide is a compound highly teratogenic to human embryos. Its teratogenic effect has been reproduced in monkeys and rabbits, the anomalies found resembling those occurring in man. The mechanism of thalidomide embryopathy has not been discovered, despite intensive research. The results of animal experiments have demonstrated their limited applicability to man and emphasize what great caution must be exercised when trying to interpret their significance for clinical medicine.

HISTORY OF MECLIZINE

It was only natural that the thalidomide experience should have focused attention on the possible teratogenic effects of several medicines used during pregnancy. Several false alarms were sounded. Short notes about the teratogenic properties of numerous drugs taken during pregnancy began to swamp the "Letters to the Editor" section in many medical journals. Their value was very limited because conclusions had been drawn from a few cases and many writers made not the slightest attempt to present any controls. Among the drugs erroneously suspected, meclizine gained a good deal of publicity. Several aspects of the meclizine story are informative, as they show the difficulties encountered in attempting to evaluate the potential teratogenicity of a drug in clinical medicine.

Meclizine (Fig. 8.12) is an antihistamine widely used to combat

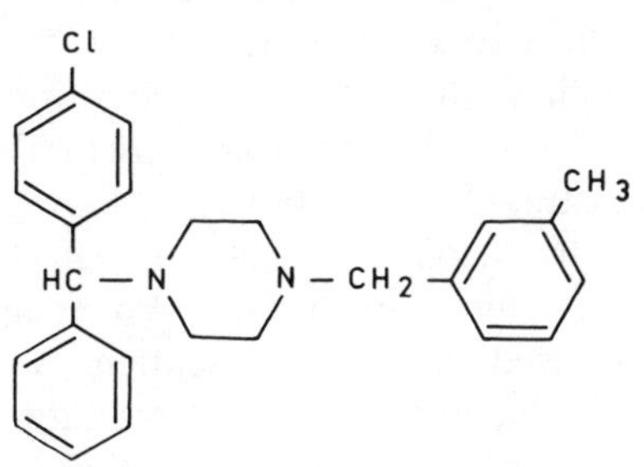

Fig. 8.12 Structural formula of meclizine.

nausea and vomiting and it has therefore been prescribed in pregnancy to mitigate the morning sickness typical of early pregnancy. Because it is also a mild sedative, the indications for its use during pregnancy overlap those of thalidomide.

On the basis of a retrospective study in which seven out of 100 women who gave birth to infants with skeletal abnormalities had used meclizine during early pregnancy and in which three out of 41 women who had given birth to infants with meningomyelocele had used the same drug, the National Board of Health in Sweden issued a warning against the use of meclizine during pregnancy (Winberg, 1963). This was followed by a number of retrospective and prospective studies on the possible teratogenic properties of meclizine. Sjövall and Ursin (1963) in Sweden showed that out of 272 patients given meclizine before month 5 of pregnancy in 1957–1962, seven gave birth to a malformed child (2.6 percent). These seven malformed children displayed a wide array of malformations with no unifying feature. In these studies the overall incidence of malformations from several years' records was 3 percent of all pregnancies.

In another study, performed prospectively at Columbia University, New York, the outcome of 3200 pregnancies was carefully studied and analyzed (Mellin and Katzenstein, 1963). Of these pregnancies, 266 (8.3 percent) resulted in malformed infants. Five of these 266 mothers had taken meclizine during the first trimester of pregnancy. In two control groups, each comprising 266 mothers of normal infants, 5 and 8 mothers, respectively, had used the same drug. It was estimated that altogether 60 mothers out of the total of 3200 had used meclizine and five of them had a malformed child. This is 8.3 percent, which is the same proportion as the overall incidence of malformations in this series.

When a drug is used to counteract a disease or a harmful condition in pregnant women and the teratogenicity of the treatment has to be evaluated, the effect of the condition itself has to be taken into account as well. Such a study has been performed in the case of meclizine and the main results are presented in Table 8.6.

The study illustrated in Table 8.6 shows that neither the condition itself nor its medication has any effect on the incidence of malformations. The meclizine group also compares favorably in all other parameters used in this investigation. A noteworthy finding was that gravidas not having nausea or vomiting during early pregnancy had a higher abortion rate than the others. This had also been suggested earlier, but the reasons are unclear and fall beyond the scope of the present survey.

Meanwhile, studies were made using meclizine and related drugs

TABLE **8.6**

Outcome of Pregnancy and Incidence of Congenital Malformations among the Offspring of Gravidas Joining the Study Group in the First 12 Weeks of Pregnancy.[a]

Diagnosis and Therapy	Number of Births	Percentage of Abortions before 20th Gestation Week	Percentage of Perinatal Deaths	Percentage of Malformations
No nausea or vomiting	1113	10.4	4.1	7.7
Nausea and vomiting	3543			
With medication:				
meclizine	330	3.03	3.44	6.3
all antiemetics[b]	1223	2.76	2.89	6.0
No medication	2320	4.18	3.37	7.1

[a] After J. Yerushalmy, and L. Milkovich: Amer. J. Obstet. Gynecol. **93:** 553, 1965.
[b] Includes all antiemetic medication prescribed by physicians and received "over the counter" without prescription.

with rats. King and his associates (1965) showed that meclizine and closely related antihistaminic drugs caused anomalies in a very high proportion of rat fetuses if 50–125 mg/kg was administered to pregnant rats on days 12 to 15 of gestation. The malformed fetuses displayed a typical and reproducible syndrome, including cleft palate, fusion of palate to tongue, and skeletal anomalies.

Evidence collected from clinical experience and from animal experiments with thalidomide, on the one hand, and from the meclizine history, on the other, show the extreme difficulty attending attempts to predict the potential human teratogenicity of a drug. Imagine a situation in which thalidomide and meclizine had been given to an experimental teratologist for the purpose of evaluating the potential teratogenicity of these drugs. He would probably have started work with the most widely used laboratory animals—rats and mice. If the outcome of his experiments had been the same as the results we now know for these drugs, his conclusions might have been as follows:

Thalidomide in high concentrations is embryotoxic and causes malformations in small percentages of mice and occasionally in rats, too. It is hardly likely to be harmful to the human embryo at the dosage levels used in clinical medicine. Meclizine, on the contrary, constantly causes a typical malformation syndrome in rats and if the time and dosage are suitably chosen, the incidence is 100 percent.

From these results it seems likely that meclizine might be a potential teratogen to man as well. We know how mistaken these speculative, but quite logical, conclusions would have been. It is abundantly clear that the results of animal experiments cannot be directly extrapolated to clinical medicine. Without animal experimentation, however, no progress can be made; therefore the experiments must be carefully planned and controlled and their significance interpreted with extreme caution, if the results suggest a connection with clinical medicine.

SUGGESTED READINGS

Review Articles

Cahen, R. L. Experimental and clinical chemoteratogenesis. *In* Advan. Pharmacol. (S. Garattini and P. A. Shore, eds.), vol. 4, pp. 263–349. Academic Press, New York and London (1966).

Dagg, C. P. Teratogenesis. *In* Biology of the Laboratory Mouse, (E. L. Green, ed.), pp. 309–328, McGraw-Hill, New York (1966).

Fave, A. Les embryopathies provoquées chez les mammifères, Extr. Thérapie **19**: 43–164 (1964).

Kalter, H. and J. Warkany. Experimental production of congenital malformations in mammals by metabolic procedure. Physiol. Rev. **39**: 69–115 (1959).

Robson, J. M., F. M. Sullivan, and R. L. Smith (ed.). Embryopathic activity of drugs. J. & A. Churchill Ltd., London (1965).

Special Articles

Beck, F. and J. B. Lloyd. The teratogenic effects of azo dyes. *In* Adv. Teratol. (D. H. M. Woollam, ed.), vol. 1, pp. 131–193. Logos Press, London, Academic Press, New York (1966).

Beck, F., J. B. Lloyd, and A. Griffiths. Lysosomal enzyme inhibition by trypan blue: A theory of teratogenesis, Science **157**: 1180–1182 (1967).

Delahunt, C. S. and L. J. Lassen. Thalidomide syndrome in monkeys, Science **146**: 1300–1305 (1964).

Fratta, I. D., E. B. Sigg, and K. Maiorana. Teratogenic effects of thalidomide in rabbits, rats, hamsters and mice. Toxic. Appl. Pharmacol. **7**: 268–286 (1965).

Gottschewski, G. H. M. Mammalian blastopathies due to drugs. Nature **201**: 1232–1233 (1964).

Hale, F. Pigs born without eyeballs. J. Heredity **24**: 105–106 (1933).

Heine, W., H. Kirchmair, M. Fiedler, and W. Stüwe. Thalidomid-embryopathie im Tierversuch. Z. Kinderheilk. **91**: 213–221 (1964).

Hendrickx, A. G., L. R. Axelrod, and L. D. Clayborn. Thalidomide syndrome in baboons. Nature **210**: 958–959 (1966).

Keberle, H., J. W. Faigle, H. Fritz, F. Knüsel, P. Loustalot, and K. Schmid. Theories on the mechanism of action of thalidomide. *In* Embryopathic activity of drugs (J. M. Robson, F. M. Sullivan and R. L. Smith, eds.), pp. 210–226. J. & A. Churchill Ltd., London (1965).

King, C. T. G., S. A. Weaver, and S. A. Narrod. Antihistamines and teratogenicity in the rat. J. Pharmacol. Exp. Therap. **147:** 391–398 (1965).

Kochar, D. M. and E. M. Johnson. Morphological and autoradiographic studies of cleft palate induced in rat embryos by maternal hypervitaminosis. A. J. Embryol. Exp. Morphol. **14:** 223–238 (1965).

Landauer, W. and E. M. Clark. The interaction in teratogenic activity of the two niacin analogs 3-acetylpyridine and 6-aminonicotinamide. J. Exp. Zool. **151:** 253–258 (1962).

——— On the teratogenic interaction of sulfanilamide and 3-acetylpyridine in chick development. J. Exp. Zool. **156:** 313–322 (1964).

Larsson, K. S. and H. Boström. Teratogenic action of salicylates related to the inhibition of mucopolysaccharide synthesis. Acta Paediat. Scand. **54:** 43–48 (1965).

Lucey, J. F. and R. E. Behrman. Effect upon pregnancy in the Rhesus monkey. Science **139:** 1295–1296 (1963).

Marin-Padilla, M. and V. H. Ferm. Somite necrosis and developmental malformations induced by vitamin A in the golden hamster. J. Embryol. Exp. Morphol. **13:** 1–8 (1965).

Mellin, G. W. and M. Katzenstein. Meclozine and foetal abnormalities. Lancet **1:** 222–223 (1963).

Robson, J. M., E. Poulson, and F. M. Sullivan. Pharmacological principles of teratogenesis. *In* Embryopathic activity of drugs, (J. M. Robson, F. M. Sullivan and R. L. Smith, eds.), pp. 21–35. J. & A. Churchill Ltd., London (1965).

Runner, M. N. and C. P. Dagg. Metabolic mechanisms of teratogenic agents during morphogenesis. *In* Symposium on normal and abnormal differentiation and development, (N. Kaliss, ed.), Nat. Cancer Inst. Monograph No. **2**, pp. 41–54, (1960).

Sjövall, A. and I. Ursing. A rapid retrospective analysis concerning the supposed teratogenicity of "Postafène" (Meclozine). *In* Possible teratogenic action of certain drugs used therapeutically in the pregnant woman. Brussels International Symposium, (P. Wilkin and M. Thiery, eds.), pp. 135–138, Imprimerie Medicale et Scientifique (S.A.), Bruxelles (1963).

Toivanen, A., T. Markkanen, R. Mäntyjärvi, and P. Toivanen. Microbiologically determined pantothenic acid and nicotinic acid content of chick embryos after treatment with thalidomide and of rat fetuses, newborns and placentas from mothers treated with thalidomide. Biochem. Pharmacol. **13:** 1489–1497 (1964).

Tuchmann-Duplessis, H. and L. Mercier-Parot. The teratogenic action of the antibiotic Actinomycin D. *In* Ciba Foundation Symposium on Congenital Malformations, (G. E. W. Wolstenholme and M. O'Connor, eds.), pp. 115–128. J. & A. Churchill Ltd., London (1960).

Winberg, J. Report on an attempt to evaluate the role of drugs in human malformations. *In* Possible teratogenic action of certain drugs used therapeutically in the pregnant woman, Brussels International Symposium, (P. Wilkin and M. Thiery, eds.), pp. 135–138, Imprimerie Medicale et Scientifique (S.A.), Bruxelles (1963).

9

Virus and Embryo

In 1941, the Australian ophthalmologist Gregg noticed a significant correlation between the incidence of congenital eye defects (cataract) and maternal rubella contracted during pregnancy. Twenty-five years later rubella virus was recovered from such cataractous lenses removed by operation some months after birth from children of mothers who had had rubella infection during pregnancy. These two observations are important landmarks in human teratology and show conclusively that a maternal virus disease can be transmitted to the fetus, where the virus persists and causes severe tissue damage, manifested as congenital defects. It is only natural that Gregg's observation soon led to many new investigations, both clinical and experimental, to discover whether rubella virus should be considered an exception or a typical example of a virus-embryo relationship. In other words, are the tissues of the embryo more vulnerable to virus infections than those of the mature organism and are the frequent, often subclinical or latent, virus infections of mothers responsible for some of the malformations of unknown origin? The answer was in the affirmative.

VIRUS-CELL INTERACTION

Before we come to the problem of viral susceptibility and viral lesions in embryonic tissues and cells, some general comments on virus-cell interaction should be made. How does a virus, a package of foreign nucleic acid, affect cells and what kind of interference with the hosts' metabolic machinery is to be expected? The prerequisite for a viral lesion and virus multiplication is the entry of the virus into the cell. This can take place through specific enzymatic processes at certain loci of the cell surface leading to the breakdown of the cell barrier, or, in the case of animal viruses, the whole virus particle can be swallowed by the cell through phago- or pinocytosis. During this stage, the viral nucleic acid, DNA or RNA, is divested of its protein coat and the free nucleic acid subsequently has two goals to fulfill: (1) it should replicate to form new, identical nucleic acid molecules and (2) it must build a new protein coat. The DNA viruses act here more or less like the genomic DNA of the host, utilizing the latter's enzyme system for replicating themselves and making their own messenger coding the synthesis of the protein coat. Viral RNA, on the other hand, codes its own polymerase, after which it can replicate without DNA-dependent control. This active synthesis of new proteins and replication of foreign nucleic acids may, by making use of the synthesizing machinery of the cell, have three different consequences:

1. Multiplication of the virus may rapidly interfere with the normal metabolism of the cell and soon lead to total destruction of the host. In light microscopy this is seen as vacuolation, hypertrophy, and nuclear displacement, ultimately leading to total disintegration. Electron microscopists have described early changes in ribosomes and furthermore in nucleoli, mitochondria, and cell membranes, all indicating profound interference with the synthetic activity of the cell. The final result is destruction of the cell and release of complete viruses, virions, ready to enter new cells and spread the infection.
2. Several viruses are known to be released by seemingly unaltered, viable cells in which replication of viral nucleic acid and synthesis of the viral protein apparently has not damaged the cell.
3. A third possibility, "symbiosis" of the viral nucleic acid and the host, can be envisaged. Here the virus is incorporated into the genome and will be transmitted to the progeny cells until some environmental "inductor" restores its independence. Thereupon the virus may start rapid multiplication resulting in the consequences mentioned in alternative 1 and finally in the destruction of its former symbiont.

VIRAL SUSCEPTIBILITY OF EMBRYONIC CELLS

Let us now return to the original question and search for experimental evidence of the great susceptibility of embryonic cells as compared with those in more advanced stages of development. Such changes in susceptibility have been demonstrated at both the cellular and organismal levels and it seems justifiable to speak of a "maturation resistance" as a widespread (although certainly not universal) phenomenon. When cells of rat embryos in different stages of development are cultivated *in vitro* and their capacity to support the growth of myxoma virus is determined, results are obtained which clearly show a gradual decrease of this capacity with advancing age (Fig. 9.1).

In order to analyze such age-dependent changes in viral susceptibility at the cellular level, simplified model systems are needed, where an acquired resistance could be correlated to actual developmental events. A great variety of known and unknown factors operate at the organismal level and the only way to extract a few of them for experimental exploration seems to be to employ such simplified conditions. Hence, we may briefly describe two such experiments performed in organ cultures of embryonic tissues and consider whether the conclusions drawn have any applicability at the organismal level. The secretory tubules of the mammalian kidney develop from the metanephrogenic mesenchyme when stimulated by the inductive trig-

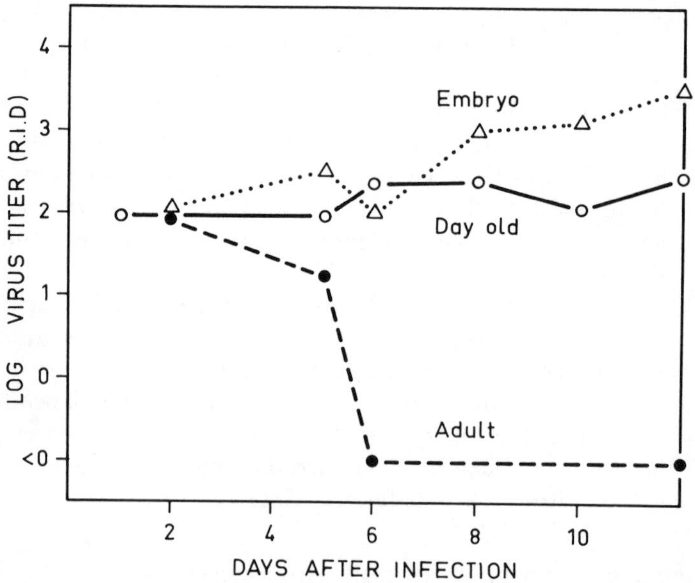

Fig. 9.1 Growth of myxoma virus in organ cultures of rat kidney tissue derived from hosts of different ages. (After D. M. Chaproniere. Virology **4:** 393 [1957].)

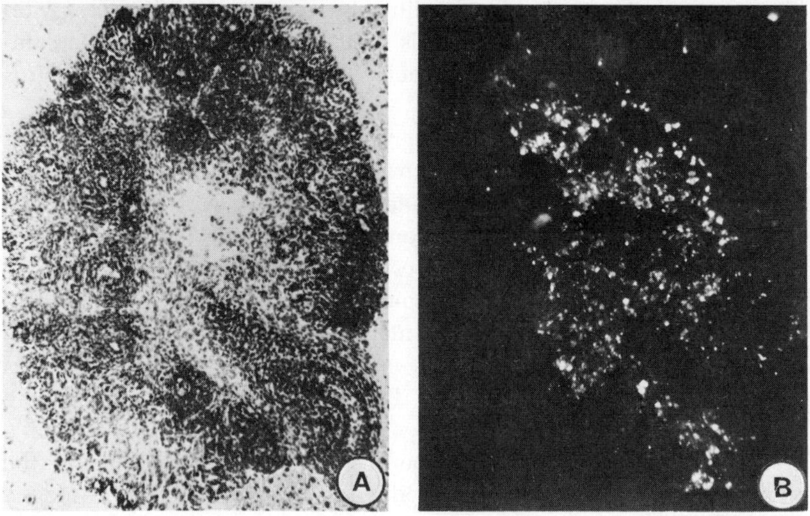

Fig. 9.2 Distribution of viral antigens in embryonic mouse kidney infected with polyoma virus *in vitro* and sectioned 5 days later. The viral antigens are demonstrated in ultraviolet light by employment of polyoma antiserum conjugated with fluorescent isothiocyanate (B). Comparison with the micrograph of the same section stained by the routine method (A) indicates that the tubules contain no viral antigens, in contrast to their progenitor cells of the loose mesenchyme. (From T. Vainio *et al.* Virology **20:** 380 [1963].)

ger action of the cells of the ureteric bud (as already described in Chapter 2). In most mammals formation of new tubules proceeds throughout intrauterine life, and simultaneously with the maturation of the oldest, centrally located tubules, new tubular condensates develop in the periphery. Consequently, the kidney rudiment of an advanced embryo consists of cells forming a developmental series from undifferentiated mesenchyme to mature elements of the tubule wall. When such rudiments are infected with various viruses the distribution of virus particles, virus antigens, and viral lesions show a rather similar pattern; the viruses seem to multiply and cause cytopathic changes only in the undifferentiated mesenchyme, indicating that the conversion of these cells to differentiated epithelial cells parallels or produces an overt change in viral susceptibility and the acquisition of a relative maturation resistance (Fig. 9.2).

Not only does normal differentiation lead to changes in the viral susceptibility of embryonic cells, but similar changes have been observed in connection with experimentally altered morphogenesis. It was found long ago and has since been repeatedly confirmed that certain embryonic epithelia can be converted from a squamous type to mucus-secreting cuboid epithelium by treatment with excess vitamin-A. A most interesting finding is that this induced metaplasia coincides with a change in the capacity of the cells to support virus growth, as seen in Fig. 9.3.

These and several other experiments *in vitro* have shown that changes in viral susceptibility occur during ontogenesis, when studied at the cellular level under simplified conditions. Not only does the capacity to support virus replication change during development, but the responsiveness and type of response to virus infection seem to be dependent on normal developmental events. Dawe and his associates (1966) have presented a model system in which the tissue response to an oncogenic virus is correlated to interactive processes between two tissue components. This interaction, normally leading to induction of normal morphogenesis, seems in this case to be a prerequisite for a neoplastic response and infection if either of the components separately fails to produce a typical oncogenic response.

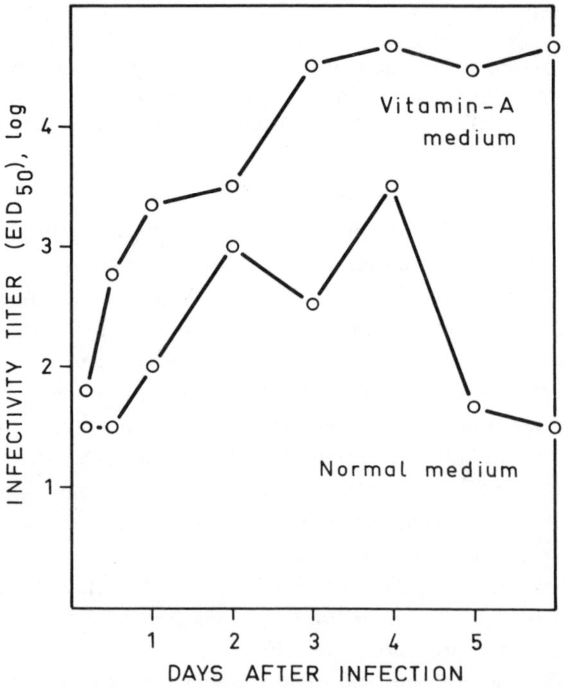

Fig. 9.3 The effect of vitamin A on a daily yield of influenza virus from chick embryo epidermis in organ culture. After 6 days of cultivation the vitamin-treated epidermis shows a transformation into cuboid epithelium, whereas advanced keratinization was noted in the control cultures. (After J. S. Chiang and F. B. Bang. J. Exp. Med. **120:** 129 [1964].)

Tissue culture studies have been helpful in analyzing the changes at the cellular level and the mechanisms possibly underlying them, but they have led us far from the original question of the role of viruses in producing malformations *in vivo*. They have indicated, however, that the problem is meaningful and that differences in reactions to viruses between embryonic and adult organisms can be expected.

MALFORMATIONS EXPERIMENTALLY INDUCED BY VIRUSES

If the conclusions based on *in vitro* studies are valid at the level of the whole organism, two general features are to be expected during ontogenesis: (1) young embryos, containing relatively high proportions of undifferentiated cells, should be more susceptible than older stages or adults and (2) a gradual restriction of susceptibility might be expected during development. Both of these phenomena have been demonstrated in chick embryos, although no generalizations should yet be made. Infection of chick embryos by different viruses during days 4 to 10 leads to fetal death in a high percentage of cases, but after day 10 the rate of survival clearly increases. The fact that the effect is really caused by the virus inoculation can be established in two ways: (1) inactivated virus preparations have no similar effects and (2) treatment of the embryos with specific antiserum prevents the lethal effect (Fig. 9.4). In these early stages viral lesions seem to be rather generalized, but during subsequent development they become restricted to certain tissues, varying according to the animal strain used and the virus inoculated. Influenza-A virus may serve here as a typical example: all tissues of the 4-day chick embryo seem to be susceptible to this virus in contrast to older embryos (13 to 15 days), in which only the respiratory epithelium responds to the infection.

True congenital defects were reported for the first time in 1947 by Hamburger and Habel, who observed a typical syndrome of maldevelopment in chick embryos inoculated with influenza-A virus at the age of 48 hours. Recently, the matter has been thoroughly examined by Williamson, Russel, and Blattner, who, in addition to analyzing influenza virus, have analyzed the effect of mumps and Newcastle disease viruses on chick embryos. All three myxoviruses caused cellular damage in the embryos, leading to several malformation syndromes, including defects in the lens, central nervous system, auditory vesicle, and visceral arches. The type and complexity of the malformations were directly related to the stage of development at the time of inoculation with the virus. When embryos infected with Newcastle disease virus were subsequently tested for viral antigens by means of fluorescent antiserum, the pattern of fluorescence

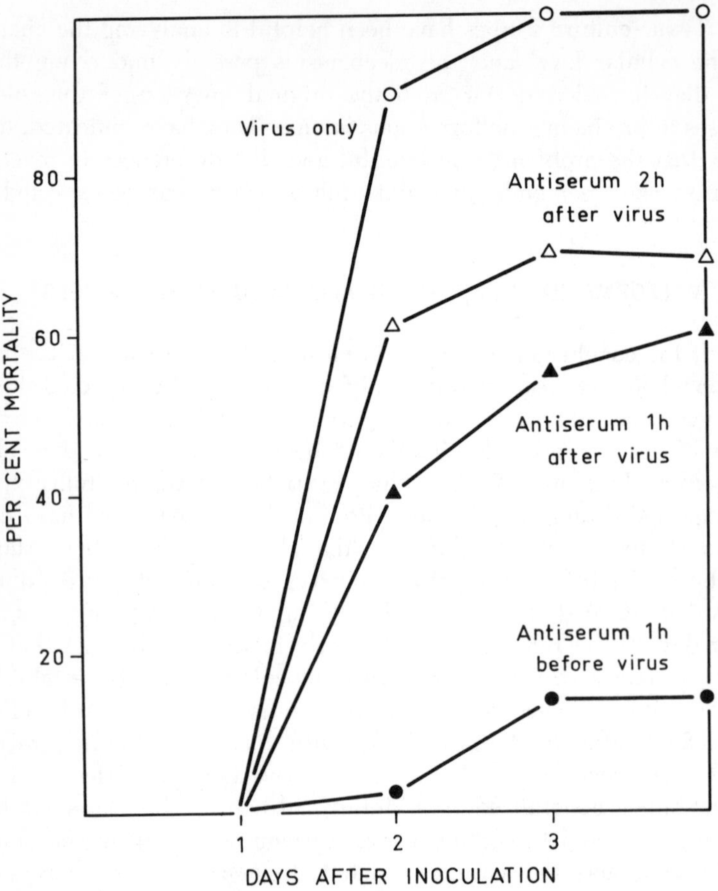

Fig. 9.4 An experiment demonstrating the role of influenza-A virus in fetal death of chick embryos: the preventive effect of antiserum. (After H. H. Shear *et al.* Proc Soc. Exp. Biol. Med. **89:** 523 [1955].)

in general corresponded to the distribution of susceptible tissues as compared to the earlier teratogenic studies.

As far as virus-induced malformations in mammals are concerned, only a few examples have been detected. Attenuated preparations of hog cholera virus cause limb malformations in the pig, and a live blue-tongue virus vaccine has been reported to be correlated to fetal deaths and epidemic abortions, indicating fetal contamination with certain diseases in cattle.

Among the few experimental approaches to the study of viral teratogenicity in mammals, the investigations of Ferm and Kilham (1964) should be mentioned. Using hamsters as test animals, these workers tested several viruses, including mumps and herpes simplex, and showed that these two did not penetrate the placental barrier. A strain of the rat virus, H-1, on the contrary, was found in both placental and fetal tissues after intravenous inoculation of the mother. Inoculation of an undiluted virus preparation on day 6 resulted in almost 100 percent resorption of the embryos. With gradual dilution of the virus, the resorption rate declined and the surviving embryos showed a variety of malformations, including exencephaly, spina bifida, and facial defects. Microscopic examination indicated intranuclear inclusions and focal necrosis in mesodermal and other tissues.

RUBELLA

As mentioned in the introduction of this chapter, rubella infection during early pregnancy has proved to be teratogenic in man; a variety of deformities and neonatal conditions have been attributed to the virus. The conclusive demonstration of this causal relationship raises several questions: At what stage of development can the embryo be damaged? What is the risk of fetal damage at different stages of pregnancy? What malformations and pathologic conditions of the newborn might be attributable to maternal rubella? And finally, what is the mode of action of the virus? The first two questions are intimately connected and should be treated together.

Sensitive Period and Risk in Rubella Embryopathy

The first reports from Australia from 1941 through 1948 indicated a very high percentage of fetal damage after maternal rubella in the first trimester—some studies suggested a percentage close to 100. Here, as always in teratology, reliable statistics are very difficult to obtain and an estimate of the fetal damage is entirely dependent on the methods used. Knowing today that even subclinical infections of the mother may lead to fetal wastage (Sever, 1967), it seems that, in the future, only virologic and immunologic studies can give a final answer to this long debated problem. Yet it may not be correct to attribute all the great differences between the results of different investigators to limitations of material or inadequate methods. The nature of the epidemic and the virulence of the virus may vary, thus causing true variations in the risk. Some of the statistics are summarized in Fig. 9.5.

One reason for the wide variations might be a difference in rates of abortion following epidemics caused by viruses differing in virulence. It is possible that a virulent virus may cause fetal damage

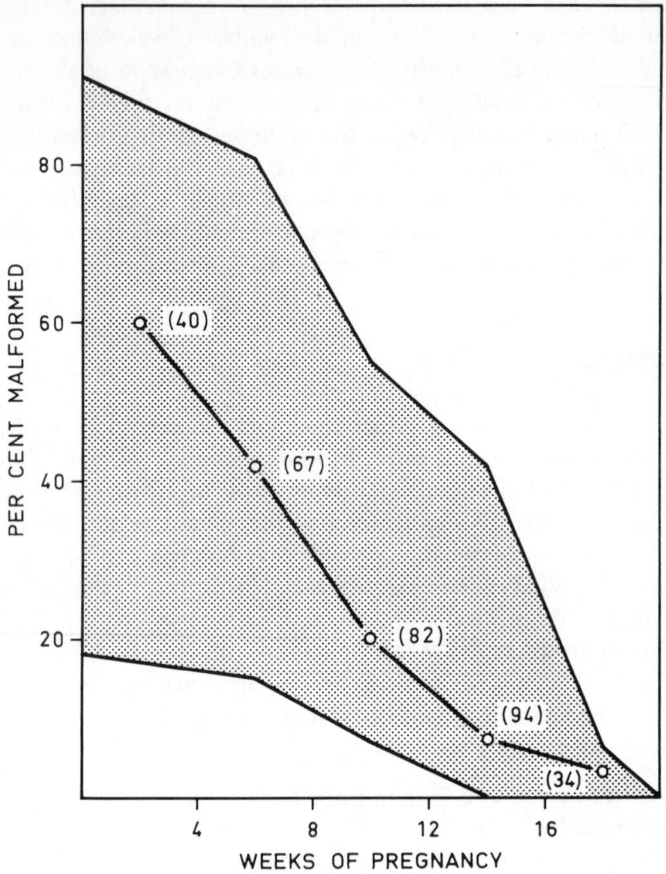

Fig. 9.5 Risk of fetal malformations following maternal rubella. The data are based on 6 independent, prospective studies summarized by E. J. Plotz. N.Y. State J. Med. **65:** 1239 [1965] and U. Baertsch. Fortsch. Geburtshilfe Gynaecol. **27:** 1 [1966]. The shaded area shows the variations between different estimates and the line shows the mean value for the risk calculated from the original data (the number of children included in each calculation is indicated).

severe enough to lead to early abortions whereas less virulent viruses cause local damage subsequently manifested as malformations. The abortion rate is naturally very difficult to estimate, but some prospective studies have indicated a figure as high as 50 percent during the first weeks of pregnancy. A detailed patho-anatomic study of embryos of mothers who contracted rubella during the first trimester has indicated a much higher percentage of damaged embryos than the statistics of liveborn infants as shown in Fig. 9.6. If this estimate, naturally based on a rather limited series (42 embryos), could be generalized, two explanations of the evident discrepancy would be possible. Either a proportion of the lesions are repaired during subsequent development and do not appear as congenital defects, or a certain number of these damaged embryos will be lost during early pregnancy and will not be included in such series as those presented in Fig. 9.5.

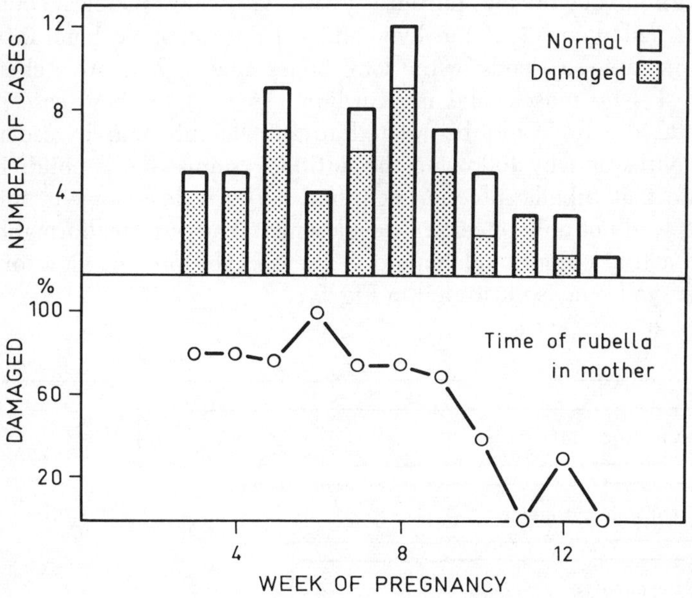

Fig. 9.6 Survey of lesions in human embryos of mothers who contracted rubella during early pregnancy. (After G. Töndury. Triangle **7:** 90 [1965].)

In conclusion, it may be stated that maternal rubella contracted during the first weeks of pregnancy constitutes a definite risk for the fetus and may lead both to early abortions and to congenital malformations. The risk may vary with different epidemics but reliable statistics on this point are very difficult to obtain. After the first trimester the embryo seems to have acquired protection against maternal rubella.

Rubella Syndrome

The original rubella syndrome consisted of three major defects: central cataract, congenital deafness, and malformation of the teeth. In addition, congenital heart diseases were frequently observed as well as some other, less regularly occurring anomalies. During 1964, a rubella epidemic was recorded in the United States and it was calculated that approximately 20,000 children were damaged. Studies of this material have yielded much new information on the rubella syndrome or "embryopathia rubeolica" (Korones *et al.*, 1965; Plotkin *et al.*, 1965; Sever, 1967).

In addition to the earlier known malformations mentioned above, several new pathologic conditions were noted that bore an evident causal relationship to the maternal disease. Among these conditions were thrombocytopenic purpura (with very low platelet counts), anemia, enlargement of the liver and spleen, osteolytic bone disease at the metaphyseal ends of the long bones, and degenerative changes in the skeletal muscle and myocardium. In most of these cases, the perinatal disease could be related to rubella infection by isolation of the virus and by following the antibody status of the child. Both indicate that rubella infection acquired early in development, might, in fact, lead not only to certain developmental arrests (malformations) but to a true chronic inflammatory disease. The present view of the rubella syndrome is illustrated in Fig. 9.7.

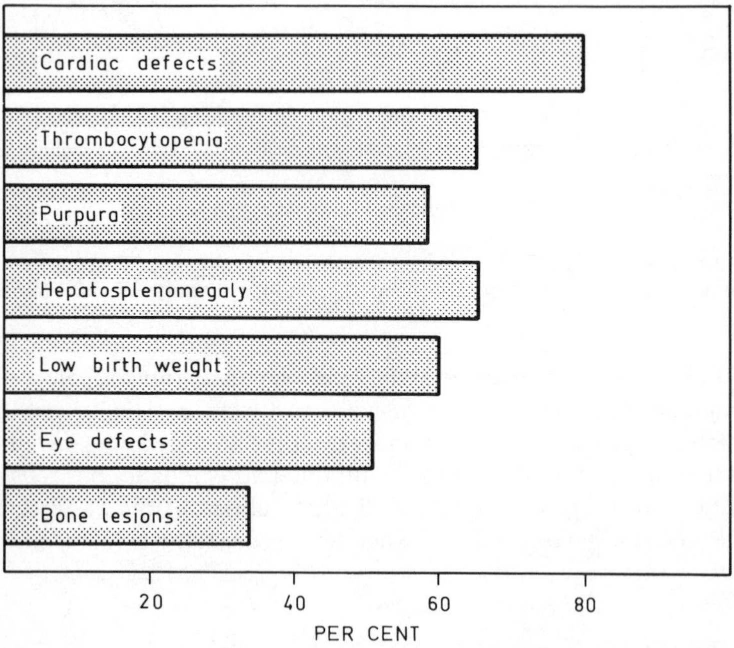

Fig. 9.7 Present concept of the rubella syndrome based on findings in 208 children with the syndrome after the rubella epidemic in the United States, 1963–1964. In 73 percent of the cases rubella virus was isolated. (Data collected by A. D. Heggie. Pediatr. Clin. North Amer. **13:** 251 [1966].)

Congenital deafness, observable only at later stages of postnatal development, is omitted from this concept and likewise certain less frequent manifestations of the disease (malformations of the teeth, microcephaly, pneumonia, myocardial lesions, degenerative changes in the skeletal muscles, palm, and sole print anomalies, and so on). Consequently, we are dealing with a syndrome in which true congenital malformations and a persistent inflammatory disease are combined. In the former group of pathologic changes we usually see only the consequences of degenerative changes or developmental arrests dating from early embryonic life (months 2 to 3 of development). It is therefore of great interest to quote some of Töndury's main findings on early embryos. He describes cell damage and necrosis not only in the chorionic epithelium, but also in the endothelial cells of the fetal blood vessels, the myocardium, the skeletal muscle, and particularly in the lens, inner ear, and tooth rudiments. In the lens, these lesions lead to the formation of central cataract as a consequence of degeneration of the central fibers. In the inner ear, it is the cochlear duct that is affected. These early changes have led the author to certain conclusions on the probable mechanism of the virus effect.

Pathogenesis of Rubella Embryopathy

From his microscopic observations on damaged embryos, Töndury concludes that the most probable route of infection is a vascular one. Diseased cells of the chorionic epithelium are spread into the fetal circulation and the changes in the endothelial cells are indicative of lesions allowing the virus to penetrate to the tissues. Here, at least two mechanisms should be taken into consideration. Rubella virus causes cell death leading to partially irreparable tissue lesions in certain organs (embryonic lens nucleus and cochlear duct). Secondly, the subsequent persistent infection seems to have an effect on organ growth, as shown in Fig. 9.8. In contrast to children in whom fetal growth retardation is due to prenatal malnutrition, the subnormal size of certain organs is due not to too little cytoplasm in the parenchymal cells but to a subnormal number of cells in the organs. This may be due to continuous loss of cells during the chronic illness, but a true indirect inhibitory action of rubella virus on cell multiplication has also been suggested.

Recent advances in the virology of rubella point the way to an experimental approach for clarification of the mechanism of rubella embryopathy. So far, the first experiments on pregnant monkeys have yielded negative results regarding the teratogenic action of the virus, although transplacental transmission has been suggested as the mechanism on the basis of active antibody production by the newborn.

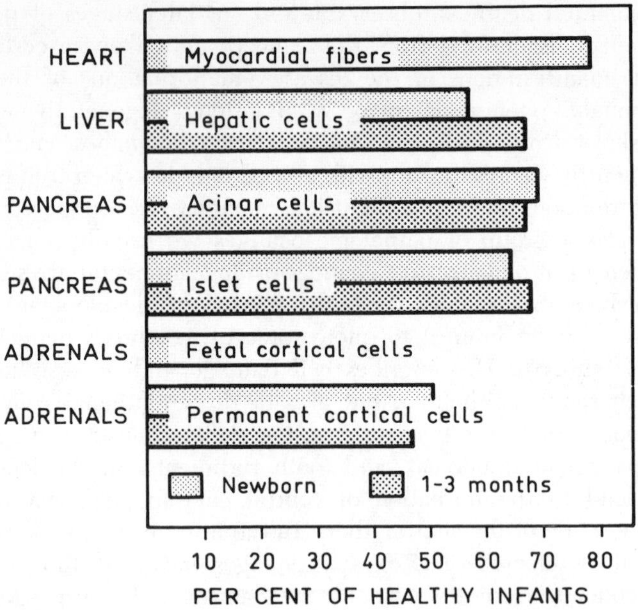

Fig. 9.8 Relative number of cells in some organs of infants with the rubella syndrome. (After R. L. Naeye and W. Blanc. J. Am. Med. Assoc **194:** 1276 [1965].)

In a recent experimental study on rats, a rubella syndrome resembling that in human beings with congenital rubella was produced (Cotlier *et al.*, 1968).

VIRUS INFECTION AS A CAUSE OF HUMAN MALDEVELOPMENT

Methods

In Chapter 2 some limitations and fallacies of epidemiologic methods were outlined; these same restrictions are also relevant to efforts to demonstrate a causal relationship between maternal viral infections and congenital malformations in human populations. Hence, the information to be gained by making use of some specific immunologic and virologic techniques has been explored. In view of the unreliability of the maternal history and the fact that even subclinical infections can cause fetal damage, such methods seem to afford the only reliable means of detecting a causal relationship, such as the one mentioned above. Three fundamentally different possibilities should be briefly mentioned: detection of viral antibodies in fetal, maternal, or neonatal serum; isolation of viruses from tissues of malformed children; and analysis of immunoglobulins of the cord blood. All these possibilities have recently been explored, with promising results.

Brown and his collaborators (see Brown, 1966) have collected blood samples from over 5000 pregnant women at their first visit to the maternity hospital (that is, prospectively). These samples were subsequently analyzed for several antibodies and finally compared to the pregnancy outcome. Several infections seemed to be more frequent among the mothers giving birth to anomalous children (for example, Coxsackie and influenza), and it is hoped that this method will yield valuable information in the future. From the fact that rubella virus can be isolated from the damaged tissues of newborn infants with a typical rubella syndrome, it might be expected that a similar correlation could be shown between other viruses and congenital defects (if teratogenic). Hence, a search has been made for viruses in cell cultures from malformed children and if extensive series can be collected, it may be possible to establish a causal relationship. So far, these investigations are still at a preliminary stage.

A third, rather recent and promising approach to the study of intrauterine infections is based on analysis of the immunoglobulins in the cord serum (Stiehm *et al.*, 1966). Such determinations have shown that only traces of the immunoglobulins gamma-M and gamma-A are normally present in the blood of a newborn infant, suggesting that these maternal globulins do not enter the fetal circulation. A definite increase in these immunoglobulins has been observed in several intrauterine infections, for example, rubella and those caused by Coxsackie virus, cytomegalovirus, and intrauterine toxoplasma. It has been suggested that this increased synthesis of immunoglobulins at a late stage of fetal development (and during the neonatal period) indicates a persistent infection such as has been described in rubella and that the method might not detect early, transient viral infections during the true sensitive period of development. In all events, a new tool seems to be available for tackling the confusing problem of the role of viral infections in congenital defects. On the following pages, other viral diseases, apart from rubella, are briefly discussed and our present, rather fragmentary knowledge is reviewed.

Influenza

Since influenza-A virus is one of the few viruses which have been shown to damage vertebrate (chick) embryos in experimental conditions, its teratogenicity in human embryos has been explored in a great number of investigations, especially subsequent to the pandemic influenza of 1957. Several epidemiologic studies have suggested a causal relationship between influenza and a slight rise in fetal mortality and in the incidence of malformations, whereas an equal number of similar investigations have failed to show such an effect (see Coffey and Jessop, 1963; Saxén *et al.*, 1960). These discrepancies might be the results of a variety of factors, among which the following should be stressed: Intense medication is used in some countries to prevent and cure the disease and some of the teratogenicity of influenza (if existing) may be attributable to the medication. Maternal memory is fallible and subclinical infections occur; in a serologic study in Baltimore, 75 percent of the women who denied having a recent flu-like infection showed serologic evidence of influenza (Hardy *et al.*, 1961). Such errors and the limitations inherent in small series may be responsible for the discrepant results rather than true differences in the nature of the epidemics and the virulence of the virus.

Smallpox and Vaccinia

A few case reports have indicated that smallpox contracted during pregnancy may harm the fetus. Because of the rarity of this disease in countries where reliable information can be collected, no conclusive data are available. But vaccination against smallpox with live vaccinia virus is known to cause transplacental infection and fetal damage (Green *et al.*, 1966). At least sixteen such cases infected at different stages of pregnancy have been reported. These cases, although few in number, show conclusively that vaccination at any stage of pregnancy may lead to infection of the fetus, with grave consequences. Hence the magnitude of the risk of these complications has to be evaluated. Several large retrospective and prospective studies have been made after mass vaccination and all but two have failed to show any significant correlation between maternal vaccination and fetal damage. The power of such epidemiologic methods has already been discussed on page 9 and as was pointed out, it is extremely difficult to exclude such fetal effect. The current view consequently is based on those isolated cases in which transplacental infection has been demonstrated and omission of vaccination is advocated during pregnancy.

Cytomegalovirus

Cytomegalic inclusion disease (CID) usually is subclinical in adults and approximately 50 percent of women of childbearing age show serologic evidence of the disease. In newborn and young infants, the virus gives rise to a severe syndrome resembling in certain features that reported in connection with maternal rubella. In many cases, infection has apparently taken place during intrauterine life, showing that the virus can cross the placenta and lead to fetal disease. The role of CID as a cause of early fetal wastage and congenital defects is less well known. Several cases of malformations, especially of the central nervous system, have been supposed to be causally related to a latent maternal cytomegalovirus infection (Medearis, 1962). Recent serologic studies have brought forward fairly convincing evidence of the causal role of this virus in the etiology of these defects, especially microcephaly. Hanshaw (1966) employed complement-fixing antibody tests in 41 children with microcephaly without other signs of CID; the test was positive in 44 percent, whereas only 4 percent of the control children gave positive serologic evidence of the disease.

Coxsackie and Echo Viruses

Several antibodies to viruses of this group have been reported to occur more frequently in the blood of mothers of abnormal children, including Coxsackie B-12, A-9, B-4, and Echo-9 (see section on Methods). Conclusive evidence of their role in the etiology of congenital defects is still lacking.

Mumps

According to Tables 9.1 and 9.2, mumps seems to be one of the most suspect teratogenic viral infections during pregnancy. In his review, Hyatt (1961) has collected from the literature 94 cases of maternal mumps during pregnancy. In 15 percent the pregnancy terminated in abortion and in another 15 percent in the birth of a malformed child. Although high, these figures may be misleading, because of the way the series was collected. There should be no doubt that the literature is biased by a strong tendency to publish cases where a positive finding has been made and not to report entirely negative findings. However, in an original series of mothers not included in the above material, Ylinen and Järvinen (1953) report an incidence of 16 percent of abortions and 23 percent of malformations after maternal mumps in the first trimester. These two independent series and the results in Tables 9.1 and 9.2 are in good agreement and suggest some causal relationship between maternal mumps and fetal defects.

Infective Hepatitis

Only scattered cases or small, inconclusive series have been reported on the effect of maternal hepatitis. An interesting suggestion has recently been made by Stoller and Collmann (1965), which, if it proves correct, seems to open up new perspectives regarding the role of viruses in abnormal development. These authors followed the annual incidence of infectious hepatitis in Australia and compared it with the incidence of Down's syndrome some 9 months later. A correlation was established, and the authors suggest that the defect was caused by infection with a virus affecting the ovum. Since several viruses are known to cause chromosomal aberrations (page 63), this claim is not implausible, but the evidence is far from conclusive and, in fact, several subsequent studies have failed to show a similar correlation in other populations.

TABLE **9.1**

Incidence of Fetal Deaths after Maternal Viral Infections during the First Trimester of Pregnancy.[a]

	Number of Mothers	Fetal Deaths (percentage)
Mumps	33	27.3
Rubella	103	20.4
Hepatitis	12	16.7
Measles	19	15.8
Chickenpox	32	15.6
Controls	1010	13.0

[a] After M. Siegel *et al.:* New Engl. J. Med. **274:** 768, 1966.

TABLE **9.2**

Incidence of Congenital Defects Following Maternal Viral Infections during Pregnancy.[a]

	Number of Mothers	Congenital Defects (percentage)
Mumps	144	13
Measles	168	7
Varicella	52	5.7
Poliomyelitis	613	1.6
Influenza	836	4.1
Infectious hepatitis	92	4.3

[a] Compiled from the literature by B. M. Kaye and B. V Reaney: Obstet. Gynecol. **19:** 618, 1962. Incidence of malformations in control series is of the order of 2 to 4 percent.

FACTORS AFFECTING VIRAL SUSCEPTIBILITY

At the beginning of this chapter, in connection with some observations on the viral susceptibility of embryonic cells, it was stressed that the conditions *in vitro* were very much simplified. At this point, after a review of some of the data on viral embryopathies in experimental animals and human embryos, the mechanism of viral diseases and factors interfering with this in embryos should be re-evaluated. This re-evaluation should be at the level of the whole organism in order to determine what conditions, genetic and environmental factors, and developmental events should be taken into consideration when investigating the problem, and which factors might be the prerequisites for viral damage during the early days of life. Only a few such factors, summarized in Fig. 9.9, can be discussed. The discussion below, however, may show the reader the extreme complication of the conditions in which we are working at the moment and how it is almost impossible to extract a single factor from the whole and expose it to the experimental exploration.

Maternal Viremia

Maternal viremia should be considered a prerequisite for viruses to enter the fetal organism and only in hypothetic cases, where the harmful effect might be transmitted by some toxic products of the infection, could an indirect effect be suspected without maternal viremia. Table 9.3 summarizes some of the most common virus diseases and our present knowledge of the presence of viremia in them.

TABLE **9.3**
Likelihood of Viremia in Man during Infection with Various Viruses.[a]

I Viremia Common	II Viremia Occurs	III Little or No Viremia
Smallpox	Mumps	Influenza
Measles	Poliomyelitis	Para-influenza
Rubella	Echo	Respiratory syncytial
Infectious hepatitis	Group B Coxsackie	Adenovirus
Yellow fever	Varicella, Zoster	Group A Coxsackie
Rift Valley fever	Attenuated measles vaccine	Herpes simplex
West Nile encephalitis	Attenuated polio vaccine	
Lymphocytic choriomeningitis	Vaccinia	

[a] Modified from R. H. Parrot: Clin. Proc. **20:** 309, 1964. Some recent findings concerning group III are not included.

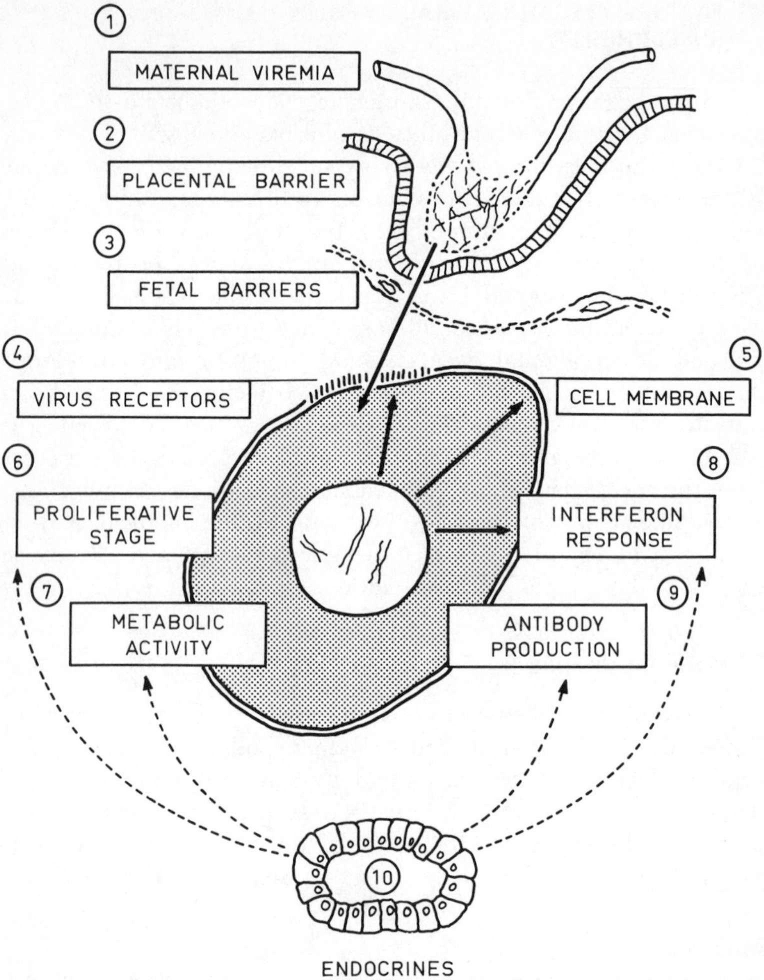

Fig. 9.9 Scheme of factors possibly affecting the viral susceptibility of an embryo.

Maternal-Fetal Barriers

Some examples already mentioned have shown conclusively that viruses can cross the placental barrier, both in laboratory animals and in man. At least the following viruses have been found in embryonic tissues (whether as manifest diseases or as isolated viruses): Japanese B encephalitis virus (swine); blue-tongue disease virus (sheep); live hog cholera virus (swine); H-1 rat virus (hamster); psittacosis-lymphogranuloma venerum-type virus (PLGV) (cattle); infectious bovine rhinotracheitis virus (cattle); rubella, herpes simplex, poliomyelitis, vaccinia, cytomegalovirus, Western equine encephalomyelitis, and Coxsackie-B viruses (human). The size of these viruses known to enter the fetal organism varies greatly; vaccinia virus is of the order of 250 to 300 mμ and the Coxsackie viruses measure only one-tenth of that. It may be concluded that size alone is not a limiting factor for viral penetration through the placenta. It should be borne in mind that several viruses (for example, rubella, poliomyelitis, varicella, cytomegalo, and vaccinia) are known to infect the placenta and may thus definitely alter the normal barrier threshold, and so promote vascular spread of the virus.

Experimental studies on transmission of viruses from the maternal circulation to the fetus are few. It was mentioned (page 217) that several rat viruses did not penetrate the maternal-fetal barrier of hamsters, whereas the H-1 variant did so. Zimmerman *et al.* (1963) used intravenous injections of Coxsackie-A9 virus to test the permeability of this barrier in rabbits and noted the first signs of virus in the blastocyst 6½ days post coitum or 6 to 8 hours before implantation, but no earlier than that. The virus could be detected in the uterine lumen throughout pregnancy, which indicated that there must be some kind of fetal barrier preventing infection before the time given above. When other workers tested this by cultivating mouse blastocysts *in vitro* and infecting them with mengo encephalitis virus, the zona pellucida—sometimes considered to be the barrier—did not prevent this small virus (27 to 28 mμ) from entering the embryo (Gwatkin, 1963). The question of corresponding fetal barriers during subsequent development is more complicated and very hard to expose to experimental testing. It has been pointed out that some of the organ anlagen (lens vesicle, auditory vesicle, and neural tube) known to be damaged by viruses are in direct connection with the surface of the embryo at the time of their susceptibility and are thus directly exposed to any virus present in the extraembryonic space. The best analyzed example is the cataract induced in chick embryos by the mumps virus, a case analyzed by Robertson *et al.* (1964). The authors could show that the lens primordium was susceptible only prior to closure and separation of the epidermis. They put forward the tempting hypothesis that the closed, infected vesicle serves as an incubation chamber, the lens capsule preventing escape of the virus from the vesicle, so that very high virus titers occur inside it (see Fig. 9-10). On the other hand, the apparent temporal relationships mentioned between susceptibility and morphogenesis might be pure coincidence and many other morphogenetic and biochemical events coincide with the change in susceptibility (for example, the initial formation of lens proteins). A similar situation will be dealt with in the next paragraph.

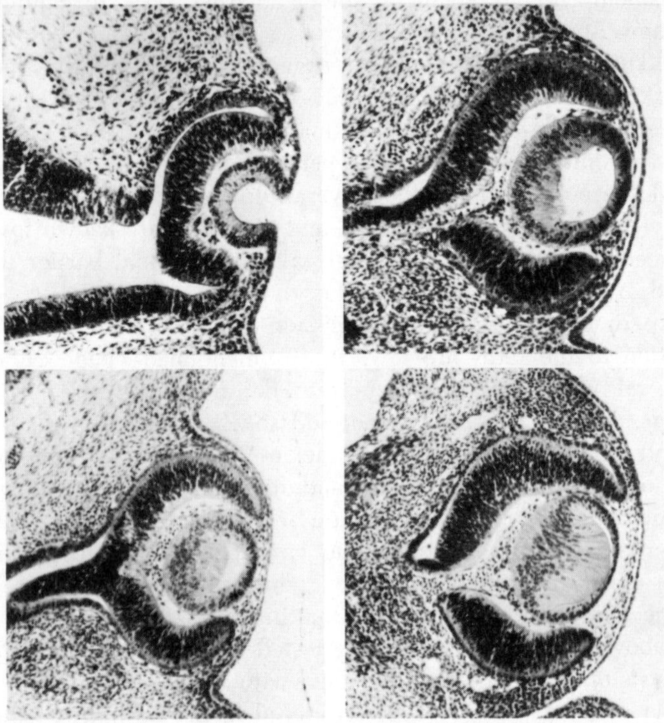

Fig. 9.10 A series of photomicrographs showing four different stages in the development of the lens in mouse embryo.

The exposed location mentioned may promote virus infection in these organs, but is certainly not a prerequisite for it, as was shown by the histologic findings of Töndury, already mentioned, and by the experimental results of Williamson *et al.* (1965). The latter authors followed the spread of Newcastle disease virus with fluorescent antiserum and noted a rapid spread into almost the entire embryo. The only exceptions seemed to be the optic cup and the brain tissue, which did not show viral antigens and, as one possible mechanism, the authors mentioned the protective effect of the basement membrane through which the virus must enter the cells (see below).

Finally, the vertical transmission of virus infection from a viremic mother to her progeny should be mentioned. In the case of avian leucosis virus (ALV), newly hatched chicks of viremic hens are known to be infected and the infection persists for long periods, perhaps throughout life, eventually manifesting itself in the form of a malignant disease. The fact that no circulating antibodies are found in these birds suggests a congenitally acquired tolerance and, consequently, a true congenital transmission of the virus from hen to egg. Recently, DiStefano and Dougherty (1966) performed experiments in which nonviremic hens were infected with ALV, with the consequence that all the newly hatched chicks were viremic. Electron microscopy of the reproductive organs of the hens revealed definite evidence of virus multiplication and suggested that the egg becomes infected either as a germ cell in the ovary or during subsequent passage through the infected oviduct.

Cell Membranes and Virus Receptors

After it has penetrated the placental barrier and entered the tissue, either through the vascular endothelium or directly via the outer surface, the virus encounters the cell membranes. Rapid changes in viral susceptibility during embryonic life may well be attributable to changes in the permeability of the extracellular membranes (basal membrane and cell periphery), but direct evidence is lacking. In the example described earlier in this chapter, acquired resistance followed the formation of pretubular condensates in the metanephrogenic mesenchyme. Parallel changes in the cell membranes were analyzed under the electron microscope, but the basement membrane barrier could be excluded because the first signs of secretory activity leading to its formation were only subsequently noted. No submicroscopic changes were visible in the cell surface at the time of early tubule formation, but this should not be taken as evidence that penetration was unchanged. In the other example mentioned, where cells of different ages were cultivated *in vitro* and showed increasing resistance, the author could exclude changes in permeability (page 212).

Virus receptors are known to alter viral susceptibility of cells in certain experimental conditions, but direct evidence of such changes during ontogenesis is lacking. It has been shown that experimental procedures which greatly disturb the normal cell contact relationship may lead to receptor acquisition and virus susceptibility. It could be speculated that a reverse process takes place during embryogenesis, where the establishment of close cellular contacts and condensation of the differentiating cells seem to play a central role.

The Proliferative Stage

The proliferative stage of embryonic cells has often been mentioned in connection with different environmental teratogens and there has been a general feeling that tissues undergoing rapid cell multiplication might be more susceptible to teratogenic agents than tissues of low mitotic activity. A group of viruses attacking and destroying mitotic cells has, in fact, been described; this group includes the H-1 virus already mentioned, the Kilham rat virus (RV), and a recently isolated feline ataxia virus (FAV) (Margolis *et al.*, 1967). Both H-1 and RV viruses are known to have both embryocidal and teratogenic properties and their destructive effect is exerted specifically on tissues undergoing rapid proliferation. It is of interest to note that the sensitive period for these viruses extends beyond the period of major organogenesis (Chapter 5), for they still can cause destruction of the cerebellum during late pregnancy. This prolonged susceptibility, resembling that observed after irradiation, may find its explanation in the high mitotic activity still continuing in these parts of the central nervous system during the late stages of intrauterine life. No such correlation has been shown in the experimental studies on chick embryos infected with different viruses. The matter has been rendered still more complicated by findings in rabbit papillomas induced by Shope virus (Shope, 1962). This virus protein is synthesized only by differentiating, keratinizing cells but not by the rapidly proliferating cells of the basal layer. In fact, the synthesis of virus protein in this case seems to be incompatible with cell division.

Metabolic Activity

Changes in the viral susceptibility of embryonic cells either *in vivo* or *in vitro* have been correlated to parallel alterations in metabolic activity. The latter has been used either as an expression of the "general metabolic activity" of cells or to indicate some specific changes in its enzymatic or synthetic activities. Such correlations are very easy to find and it may be said that at any given stage of cytodifferentiation, the metabolism of a cell is quantitatively or qualitatively altered. Thus a change in viral susceptibility can certainly be temporally correlated to a variety of different changes in the biochemistry of the responding cells, but so far it has not been possible to establish a causal relationship.

Interferons

When dealing with virus infections of embryonic cells, certain antiviral products of the cells themselves and of the whole organism should be mentioned. Of these, the interferons are perhaps the best known and their suppressive action on virus infection has been repeatedly demonstrated. If interferon production is regarded as a specific synthetic response to exogenous stimuli, it is quite feasible to postulate that this capacity develops during ontogenesis and might be lacking in undifferentiated tissues. In fact, Baron and Isaacs (1961) were able to show that in chick embryos both production of interferon and sensitivity to its antiviral action increased with advancing embryogenesis and that these changes were closely correlated to viral susceptibility as measured by the survival rate. As the authors have suggested, the inability to produce interferons and insensitiveness to its antiviral action may play an important role in the high susceptibility of young embryos demonstrated in a variety of experiments. Information on interferon production by mammalian embryos is scanty. Recently, Cantell *et al.* (1968) analyzed the interferon production of human leukocytes *in vitro* and demonstrated that even the first lymphocytes entering the peripheral blood during week 9 of development produced interferon in quantities comparable to those of adult cells. It may thus be concluded that at the end of the first trimester there are interferon-producing elements in the peripheral blood, possibly contributing to the barrier excluding maternal viruses from the fetal circulation.

Maturation of the Immune System

Another defense mechanism of the embryo itself, the production of specific antibodies, must be considered responsible for the development of viral resistance. The recent findings of nonmaternal immunoglobulins in cord serum indicate that the fetus is immunologically at least partially competent at birth. It is very difficult, however, to determine the point at which the fetus starts to produce antibodies against foreign antigens and only very few observations have been reported. Human and animal embryos have been shown to produce humoral antibodies at a relatively early stage of development, but their quality and quantity are dependent on the stage of maturation. Sheep embryos will produce antibodies against certain bacteriophages as early as day 35 (total gestation period 150 days) but are still too immature to synthesize antiferritin antibodies by day 66 and certain antibacterial antibodies are not produced until 6 weeks after birth (Silverstein, 1964). Experiments on the common laboratory rodents have indicated that they do not develop an immune reaction against transplantation antigens during the intrauterine period and still accept a homologous graft shortly after birth. These observations led to the generalization that embryos in general are not capable of developing a cell-bound immune reaction during the intrauterine period. However, transplantation experiments on sheep and monkeys have revealed exceptions to this rule; the former will already reject a homologous graft at the beginning of the second half of pregnancy.

In the monkey, this immune reaction may develop even earlier (Sterzl and Silverstein, 1967). On the other hand, it has been shown that certain tumor viruses induce neoplasms only if injected during the immediate postnatal period and that the length of this susceptible period can be correlated with the time of maturation of the immune system in different strains of experimental animals. If a highly resistant mouse strain is employed in which the susceptible period is apparently confined to the intrauterine period, its susceptibility can be restored by early thymectomy and resistance can be conferred once more by inoculation of syngeneic, immunologically competent spleen cells (Law, 1965). This example strongly suggests that the oncogenic response to viruses is immunologically controlled and that in this particular case the immunologic system does not mature until the time of birth.

Although general conclusions may still be premature, the results so far indicate rather late maturation of the immune system in the fetus and this defense mechanism may not explain the acquired viral resistance usually observable during early development. However, new, sensitive methods for detecting the antibodies mentioned previously may soon change our concepts and lead to a re-evaluation of the significance of immune mechanisms during early embryogenesis.

Other Factors

A great many other factors during embryonic life may affect viral susceptibility, but only a few endocrine aspects will be mentioned here. Several hormones are known to penetrate the placenta and the endocrine activity of embryonic glands starts relatively early during ontogenesis (see Chapter 6). Hence the possible role of the hormonal status in the responsiveness of embryonic tissues to viral infections should be mentioned briefly. The importance of endocrine factors for virus susceptibility-resistance and the extreme complexity of the problem may best be exemplified by listing some of the characteristics of cortisone. The hormone is known to have drastic effects on both the metabolism and proliferation of tissues, being itself an experimentally proven teratogen (page 20). Hence it may well affect viral susceptibility, which is supposed to be partially dependent on these tissue conditions. Moreover, cortisone is known to interfere with both antibody production and the interferon response, having shown a clear inhibitory effect on both, at least in some studies. The depressive effect on the interferon response does not seem to be associated with viral replication or with a change in the antiviral activity of interferon, but it reflects a true inhibition of its production (Mendelson and Glasgow, 1966). In addition, *in vivo* experiments have repeatedly shown the sensitizing effect of cortisone, for example, on experimental poliomyelitis infection. The problem is rendered even more confused by the fact that corticosteroids have been shown to protect some cells from the cytopathic effect of certain viruses (probably by inhibiting the release of lysosomal enzymes triggered by virus infection).

VIRAL LESIONS

Finally, a few comments should be made on the lesions caused by viruses. Naturally, the final response depends, not only on the many factors mentioned above, but also on the genetic constitution of the embryonic cells and on the nature of the virus. Not only do different viruses cause quite different types of tissue damage, but certain teratologic results have indicated that the virulence of a virus may change and that teratogenic responses during different epidemics may vary. Moreover, the tissue reaction to certain viruses seems to become modified during embryogenesis. Rous sarcoma virus produces a hemorrhagic disease in young embryos, but with increasing age the oncogenic properties of the virus are manifested while hemorrhagic lesions become rare. Similarly, polyoma virus in organ cultures shows cytopathic effects on undifferentiated cells, but may lead to proliferation of cells in more advanced stages of morphogenesis.

The great variability of reactions makes any generalizations useless and the essential question is that of the mechanism of malformation due to tissue damage. However, there is one respect in which viral infection differs from all other teratogens. Virus brings new information to the infected cells which may remain there incorporated with the host genome or as an epigenetic unit. Recovery of virus from such cells and tissues may not be possible and, hence, our present views on the role of latent viruses in development are entirely speculative.

SUGGESTED READINGS

Review Papers

Blattner, R. J. and F. M. Heys. Role of viruses in the etiology of congenital malformations. Progr. Med. Virol. **3**: 311–362 (1961).

Brown, G. C. Recent advances in the viral aetiology of congenital anomalies. *In* Advances in Teratology, (D. H. M. Woollam, ed.), Vol. 1, pp. 55–75. Logos Press, London and Academic Press, New York (1966).

Ebert, J. D. and F. H. Wilt. Animal viruses and embryos. Quart. Rev. Biol. **35**: 261–312 (1960).

Hardy, J. M. B. Viral infection in pregnancy. Am. J. Obstet. Gynecol. **93**: 1052 (1965).

Heggie, A. D. Rubella: Current concepts in epidemiology and teratology. Pediat. Clin. North Am. **13**: 251–266 (1966).

Medearis, D. N. Viral infections during pregnancy and abnormal human development. Am. J. Obstet. Gynecol. **90**: 1140–1148 (1964).

Rhodes, A. J. Virus infections and congenital malformations. *In* Congenital Malformations (M. Fishbein, ed.), pp. 106–117. J. B. Lippincott Co., Philadelphia (1961).

Saxén, L. Ontogenesis of virus resistance. *In* Symposium on Virus and Cancer, (H. Bergstrand and K. E. Hellström, eds.), pp. 150–161. Tryckeri Balder AB, Stockholm (1965).

Sever, J. L. Rubella as a teratogen. *In* Advances in Teratology. (D. H. M. Woollam, ed.), vol. II, pp. 127–138. Logos Press, Acad. Press, London (1967).

Sterzl, J. and A. M. Silverstein. Developmental aspects of immunity. Advan. Immunol (F. J. Dixon and J. H. Humphrey, eds.), **6** pp. 337–359, Academic Press, New York and London (1967).

Special Articles

Baron, S. and A. Isaacs. Mechanism of recovery from viral infection in the chick embryo. Nature **191**: 97–98 (1961).

Cantell, K., H. Strander, L. Saxén, and B. Meyer. Interferon response of human leukocytes during intrauterine and postnatal life. J. Immunol. **100**: 1304–1309 (1968).

Coffey, V. P. and W. I. E. Jessop. Maternal influenza and congenital deformities. Lancet **1**: 748–751 (1963).

Cotlier, E., J. Fox, G. Bohigan, C. Beaty, and A. DuPree. Pathogenic effects of rubella virus on embryos and newborn rats. Nature **217**: 38–40 (1968).

Dawe, J. D., W. D. Morgan, and M. S. Slatick. Influence of epitheliomesenchymal interactions on tumor induction by polyoma virus. Intern. J. Cancer **1**: 419–450 (1966).

DiStefano, H. S. and R. M. Dougherty. Mechanisms for congenital transmission of avian leukosis virus. J. Nat. Cancer Inst. **37**: 869–883 (1966).

Ferm, V. H. and L. Kilham. Congenital anomalies induced in hamster embryos with H-1 virus. Science **145**: 510–511 (1964).

Green, D. M., S. M. Reid, and K. Rhaney. Generalized vaccinia in the human foetus. Lancet **1**: 1269–1270 (1966).

Gwatkin, R. B. L. Effect of viruses on early mammalian development. I. Action of mengo encephalitis virus on mouse ova cultivated *in vitro*. Proc. Nat. Acad. Sci. **50**: 576–581 (1963).

Hanshaw, J. B. Cytomegalovirus complement-fixing antibody in microcephaly. New Engl. J. Med. **275**: 476–479 (1966).

Hardy, J. M. B., E. N. Azarowicz, A. Mammini, D. N. Medearis, and R. E. Cooke. The effect of Asian influenza on the outcome of pregnancy. Baltimore 1957–1958. Am. J. Public Health **51**: 1182–1188 (1961).

Hyatt, H. W. Relationship of maternal mumps to congenital defects and fetal deaths and to maternal morbidity and mortality. Am. Practitioner Dig. Treat. **12**: 359–363 (1961).

Korones, S. B., L. E. Ainger, G. R. G. Monif, J. Roane, J. L. Sever, and F. Fuste. Congenital rubella syndrome: New clinical aspects with recovery of virus from affected infants. J. Pediat. **67**: 166–181 (1965).

Law, L. W. Thymus: Role in resistance to polyoma virus oncogenesis. Science **147**: 164–165 (1965).

Margolis, G., L. Kilham, and J. Davenport. A model for virus induced reproductive failure: Theory, observations and speculations. *In* Comparative Aspects of Reproductive Failure, (K. Benirschke, ed.), pp. 350–360. Springer-Verlag, New York (1967).

Medearis, D. N. Cytomegalic inclusion disease as an example of a viral infection acquired *in utero* which may result in mental retardation. Ment. Retard. **39**: 130–140 (1962).

Mendelson, J. and L. A. Glasgow. The *in vitro* and *in vivo* effects of cortisol on interferon production and action. J. Immunol. **96**: 345–352 (1966).

Plotkin, S. A., F. A. Oski, E. M. Hartnett, A. R. Hervada, S. Friedman, and J. Gowing. Some recently recognized manifestations of the rubella syndrome. J. Pediat. **67**: 182–191 (1965).

Robertson, G. G., A. P. Williamson, and R. J. Blattner. A study of abnormalities in early chick embryos inoculated with Newcastle disease virus. J. Exp. Zool. **129**: 5–44 (1955).

Robertson, G. G., A. P. Williamson, and R. J. Blattner. Origin and development of lens cataracts in mumps-infected chick embryos. J. Anat. **115**: 473–486 (1964).

Saxén, L., L. Hjelt, J. Sjöstedt, J. Hakosalo, and H. Hakosalo. Asian influenza during pregnancy and congenital malformations. Acta Pathol. Microbiol. Scand. **49**: 114–126 (1960).

Shope, R. E. Are animal tumor viruses always virus-like? J. Gen. Physiol. **45**: 143 (1962).

Silverstein, A. M. Ontogeny of immune response. Science **144**: 1423 (1964).

Stiehm, E. R., A. J. Ammann, and J. D. Cherry. Elevated cord macroglobulins in the diagnosis of intrauterine infections. New Engl. J. Med. **275**: 971–977 (1966).

Stoller, A. and R. D. Collmann. Incidence of infective hepatitis followed by Down's syndrome nine months later. Lancet **2**: 1221–1223 (1965).

Tartakow, I. J. The teratogenicity of maternal rubella. J. Pediat. **66**: 380–391 (1965).

Töndury, G. and D. W. Smith. Fetal rubella pathology. J. Pediat. **68**: 867–879 (1966).

Williamson, A. P., R. J. Blattner, and G. G. Robertson. The relationship of viral antigen to virus-induced defects in chick embryos. Newcastle disease virus. Develop. Biol. **12**: 498–519 (1965).

Ylinen, O. and P. A. Järvinen. Parotitis during pregnancy. Acta Obstet. Gynecol. Scand. **32**: 121–132 (1953).

Zimmermann, W., G. H. M. Gottschewski, H. Flamm, und C. Kunz. Experimentelle Untersuchungen über die Aufnahme von Eiweiss. Viren und Bakterien während der Embryogenese des Kaninchens. Develop. Biol. **6**: 233–249 (1963).

Index

Abortions, incidence of defects, 4
3-Acetylpyrimidine, teratogenic action in chick, 190–191
Acquired tolerance in virus-infected chick, 230
ACTH in adrenogenital syndrome, 152
Actinomycin D, and chromosomal aberrations, 62
  effect on kidney tubule differentiation, 127, 197
  effect on nucleic acid metabolism, 196, 197
  effect on pancreatic enzymes, 133
  sensitive period, 119, 127, 198
Additive effects of teratogens, 84
Adrenal cortex, fetal function, 142–143, 152
Adrenal gland, hyperplasia, 152
  hypoplasia in anencephaly, 142
Adrenogenital syndrome, 43, 152–154
Albinism, 42, 43
Aldosterone in adrenogenital syndrome, 154
Alizarin technique for skeletal malformations, 24, 25
Alkylating agents, and chromosomal aberrations, 62
  as teratogens, 183, 184
Amelia, 199
6-Aminonicotinamide, combined action with sulfanilamide, 189
  effect on different inbred strains, 191
  induction of micromelia, 54–55
  induction of phenocopies, 53–55
  interference with 3-acetylpyrimidine, 190–191
Amphibian metamorphosis, hormonal regulation, 144
Anaphase lagging, 58
Androgenic hormones, effect on sex differentiation, 146
  in adrenogenital syndrome, 152
Anencephalia, and adrenal hypoplasia, 142
  epidemiology, 11, 12
  and estrogenic hormones, 142
  induced by vitamin A in hamster, 194–196
Anencephalia (*Continued*)
  seasonal variations, 12
Aneuploidy, 58
Animal experiments, basic design, 19–23
  disadvantages, 27
Aniridia, association with nephroblastoma, 103
Anophthalmia, as a consequence of inductive failure, 88–89
  induced by triethylene-melamine in newt larve, 194, 195
Antibodies, against fetal antigens, 73–74
  production in sheep fetus, 233
  viral, detection in maternal blood, 223
Antigen(s), ABO, 73
  dissimilarity between mother and fetus, 72
  histocompatibility, 72
  Rhesus, 73
Asian influenza, 15, 18, 224
Atom bomb, fetal consequences, 14, 178
  relation to microcephaly, 178
Autosomal trisomy, 66–67
Autopsy series, incidence of defects, 5
Avian leucosis virus, transmission to offspring, 230

Bang, F. B., 214
Barr body, 71
Barriers, in teratogenesis, 130–131
  and acquisition of virus resistance, 228–231
Beck, F., 186
Behavior and prenatal radiation, 176–177
Bender, M. A., 62
Bennett, D., 95
Blastocyst, examination, 26
  viral infectivity in rabbit, 229
  viral infectivity of mouse *in vitro*, 229
Blastopathies, and double monsters, 117
  and teratomas, 118
Bongiovanni, A. M., 152

Brachydactyly in man, 36
in mutant rabbits, 105
Brent, R. L., 26
Bromodeoxyuridine (BUDR), and chromosomal abnormalities, 64
Brown, G. C., 223
Browne, D. A., 136
Busulfan, and congenital sterility, 131–132
Butcher, R. L., 115–116

Cardia bifida, 82
Cardiac mesoderm, effect of chelating agents, 81
migration, 79–80
Carr, D. H., 57
Cell contacts, in formation of heart, 80
role of calcium, 81
in sea urchin gastrulation, 79
Cell culture, 27
synchronous cultures, 167–168
Cell cycle, of early zygote and effect of radiation, 174–175
stages of, 167
susceptibility of different stages to radiation, 167–169
Cell death and morphogenesis, 104–105
Central nervous system, behavioral development, 25
covert defects, 25
effect of radiation, 173
epidemiology of defects, 11–13
(*see also* specific defects)
Cerebral palsy, epidemiologic studies, 9
Chaproniere, D. M., 212
Chase, E. B., 88
Chase, H. B., 88
Chelating agents, effect on cardiac mesoderm, 81
Chemical teratogens, *see* Drugs
Chemical factors and chromosomal abnormalities, 62
Chiang, J. S., 214
Chromatid breaks, 59
Chromosomal abnormalities, autosomal mosaicism, 69
autosomal trisomy, 66–67
caused by radiation in early zygote, 174–176
and chemical factors, 62–64
and delayed fertilization, 115–116
deletion of chromosome material, 68
in Down's syndrome, 66–67
epidemiology, 59
in fertilized eggs, 57
frequency, 57

Chromosomal abnormalities (*Continued*)
in human abortions, 57, 66
human autosomal, 66–67
and maternal age, 61
monosomy, 68
numerical aberrations, 58
pulverization, 65
and radiation, 61–62
of sex chromosomes, 69–71
sex chromosome mosaicism, 71
structural abnormalities, 59–60
and viruses, 63, 226
Chromosomes, aberration, 57–71
aneuploidy, 58
breaks, 59, 65
effect of radiation, 61–62
euploidy, 58
hyperploidy, 58
hypoploidy, 58
isochromosomes, 59
mosaicism, 58, 69
polyploidy, 58
ring chromosomes, 59
Clark, E. M., 189
Cleft palate, caused by cortisone in combination with other teratogens, 188
Fraser's hypothesis, 82–84
hereditary risk, 48
incidence, 47
induction by cortisone in different mouse strains, 20
induction by fluorouracil, 52
induction by meclizine, 207
and inhibition of mucopolysaccharide synthesis, 198
sensitive periods, 126
studies *in vitro*, 29
Colchicine and chromosomal aberrations, 63
Competence, *see* Inductive tissue interactions
Cataract in congenital rubella, 60, 219
Congenital defects, as causes of death,
definition, 2
detection rate, 3
incidence, 3
lost cases, 8
multifactorial etiology, 46
in registers, 8
Congenital goiter, 149
Congenital hypothyroidism, 149–150
Congenital nephrosis, 37, 38
Congenital tumors, 102
Cortisol, in adrenogenital syndrome, 152, 153
biosynthesis of, 153

Cortisol (*Continued*)
effect on palatal closure *in vitro,* 85
in mammary gland development, 147–148
Cortisone, effect on mucopolysaccharide synthesis, 198
relation to viral susceptibility, 234
synergistic action with fasting, 188
teratogenicity in different mouse strains, 193
Coxsackie viruses, 225
Cri-du-chat syndrome, 69
Cyclopy, experimental production of, 87
Cytomegalovirus, effect on human embryo, 224

Dagg, C. P., 52–53, 125, 188
Dawe, C. D., 214
Degeneration, and brachydactyly in rabbits, 104
and congenital defects, 104–108
effect of insulin on chick embryo limb, 106
excessive, 106
failure of, 106–108
failure to degenerate in mutant chick limb, 107
failure to degenerate of esophagus in crooked neck dwarf chick, 107
in formation of avian limbs, 104–105
inhibition by Janus green, 108
of inner ear in walzer-shaker mouse, 105
normal, 104
of ventral neural tube in mutant mouse, 105
DeHaan, R. L., 79–82
Dehydroepiandrosterone, synthesis in fetal adrenal cortex, 142, 143
Delayed fertilization, 115–116
and Down's syndrome, 61
5-deoxycytidine, prevention of anomalies with thymidine, 190
Diabetes mellitus, concept of pathogenesis, 158–160
insulin in treatment, 161
Diabetic mother, and fetal malformations, 157, 159
insulin in cord blood, 156, 158
and perinatal death, 156
Diazo dyes as teratogens, 183–184
Dimethyl sulfoxide and head malformations, 123
Disaggregation of embryonic cells, 129
Dorn, H. F., 16
Dose-dependence of sensitive periods, 125, 134
Double monster, 117
Down's syndrome, association with leukemia, 103
and chromosomal aberrations, 66–67
and delayed fertilization, 61
epidemiology, 59
and hepatitis epidemics, 226
incidence, 66
relation to maternal age, 61
translocation of G material, 67, 68
Drugs, additive effect, 188, 189
combined action, 187–191
dose dependence of teratogenic action, 22
effect of genotype on teratogenicity, 191–193
harmful effects during late pregnancy, 136
indirect effect on embryo, 23
interference in teratogenicity, 188, 190
maternal toxicity, 23
mechanism of teratogenic action, 193–199
mode of administration, 22
nil effect, 188, 189
potentiating effect, 188
specificity of teratogenic action, 186–187
teratogenic activity at subthreshold levels, 189
teratogenicity in man, 183
Dunn, L. C., 48, 95

Echo viruses, 225
Ede, D. A., 107
Embryonal rests and neoplasms, 3
Embryopathia rubeolica, *see* Rubella embryopathy
Enzyme(s), in adrenogenital syndrome, 152–154
in cortisol biosynthesis, 152–154
in fetoplacental unit, 142, 143
in inborn errors of metabolism, 44
lactate dehydrogenase and kidney tubule morphogenesis, 127
lysosomal, in rat yolk sac, 186
Epidemiologic studies, control material, 10
detection rate, 8
methods, 7–19
null hypothesis, 9
power of the method, 9
Epigenetic crises, 112, 126
Erythroblastosis fetalis, 73–74
Esophagus, occlusion, 107
reopening, 107

Estrogenic hormones, in anencephaly, 142
production by fetoplacental unit, 142, 143
as sign of fetal condition, 142
in Turner's syndrome, 155
Euploidy, 58
Exogastrulation, 79–80
Exostoses, multiple, 99
Experimental animal, avian embryos, 20
lower vertebrates, 20
mammals, 20
Experimental procedure, sources of errors, 20, 27
Expressivity of genes, 37–41, 56

Fasting, teratogenic effect on mouse, 189
Feline ataxia virus, 232
Ferm, V. H., 26
Fertilization, delayed, 115–116
Fetopathies, 134–137
Feto-placental unit, 141–144
Flat-mount technique for blastocysts, 26, 118
Fluorodeoxyuridine and chromosomal aberrations, 64
5-Fluorouracil, action on mutant mice, 52
and differences of sensitive periods in mouse strains, 126
induction of cleft palate, 52
induction of hemimelia, 52
induction of polydactyly, 53
Fraser, F. C., 22, 48, 82–84
Fugo, H. W., 115–116

Galactoflavin, teratogenicity in different mouse strains, 193
Gallien, L. G., 145
Gametopathies, 114–116
Garrod, A., 41
Gastrulation, disturbances, 79, 80
Gene(s), activation, 44
dominant modifiers, 50
expression, 36, 49
expressivity, 37, 56
interaction, 47–50
interplay with exogenous factors, 50–57
markers, 40
modifying, 48
mutant *see* Mutant
penetrance, 40
product, 40
regulative, 44, 45
selection of, 74

Gene(s) (*Continued*)
in sex chromosomes, 69–71
structural, 44
Genetic polymorphism, advantages of, 74
Geographic isolates, 37
German, J., 61
German measles, *see* Rubella
Gesenius, H., 13
Gluecksohn-Waelsh, S., 89
Glycosuria, renal, 46
Goiter, congenital, 149–150
Goldstein, M., 191
Gonad(s), effect on sexual differentiation, 145–146, 152
failure of development, 154
in Turner's syndrome, 155
Gonadal dysgenesis, 155
Gooch, P. C., 62
Gregg, A., 260
Grobstein, C., 31, 85
Growth, controlling factors, 97
misplaced, 99–101
neoplastic, 102
overgrowth, 99
Growth inhibition, in congenital hydrocephaly of mouse, 99
in congenital rubella syndrome, 221
and development of heart, 98
Grüneberg, H., 48, 49, 99

Hale, F., 182
Hamburger, V., 94, 215
Heart, cardia bifida, 81–82
septal defects, 98
tetralogy of Fallot, 100–101
Hemimelia, induction by fluorouracil, 51–52
in mutants *lx* and *lu*, 51–52
Hemoglobin synthesis, inhibition, 123
Hemoglobinopathies, 55
Hemophilia, 37, 39
Hemsworth, B. N., 131
Hertig, A. T., 116
Hicks, S. P., 173
Hinchliffe, J. R., 107
H-1 virus and fetal wastage, 217
Holtfreter, J., 94
Hormones, as teratogens, 161
effect on differentiation and development, 144–148
(*see also* under specific hormones)
Hydrocephaly, in *ch* mice, 99
in congenital toxoplasmosis, 135
Hydrocortisone, *see* Cortisol
3β-hydroxysteroid dehydrogenase, in adrenogenital syndrome, 154
in fetoplacental unit, 142

5-hydroxytryptamine, teratogenicity, 186
Hypophysis, dwarfism due to lack of, 149
  experimental intrauterine removal, 149
  and ontogeny of endocrine glands, 141
Hypoxia and congenital defects, 121

IgG, 73
Immune system, maturation of, 233
Immunoglobulins, in cord serum, 223
  and intrauterine viral infection, 223
  placental transfer, 73
  production by fetus, 233
Impaired learning after fetal radiation, 177
Inborn errors of metabolism, in cortisol synthesis, 152
  dominant inheritance, 45
  metabolic pathways, 41–42
  metabolism in heterozygotes, 43
  recessive inheritance, 43, 45
  in thyroid hormone synthesis, 150
Inbred strains of mouse, *see* Mouse strains
Incompatibility, ABO blood groups, 73–74
  antigens, 72
  maternal-fetal, 71–74
  Rhesus, 73
Inductive tissue interactions, 85–97
  and competence of responding tissue, 86
  in development of CNS, 86
  and ectodermal ridge, 90
  failure to make contact, 88–89
  failures of, 87–97
  and formation of cartilage, 95–96
  heterotypic stimuli, 86
  homotypic stimuli, 86
  in kidney morphogenesis, *see* Kidney
  in kidney morphogenesis of mutant mouse, 89
  in limb development, 90–94
  loss of competence, 94–97
  loss of inductive capacity, 89–94
  primary induction, 85–87
  and viral susceptibility, 213
Influenza, and congenital defects in man, 223–224
  epidemiologic studies, 15, 224
  (*see also* Asian influenza)
Influenza virus, effect on chick embryos, 215
Ingalls, T. H., 11, 12, 46
Inheritance of defects, autosomal dominant, 36
  autosomal recessive, 36
  X-chromosome-linked, 36
Inhorn, S. L., 57
Insulin, activity in cord blood, 156, 158
  as a cause on chick limb malformations, 106
  and congenital defects, 161
  in mammary gland development, 147–148
Interaction(s), of genes, 47–50
  of genes and exogenous factors, 50–57
  inductive, 85–97
  of trophoblast and maternal endometrium, 73
Interferon, in chick embryos, 233
  production in human fetal leukocytes, 233
Intrauterine experiments, in analysis of endocrine development, 141
  in analysis of sex differentiation, 145
Intrauterine manipulations, hypophysectomy, 141, 149
  irradiation, 26
  transplantation, 26
  vascular clamping, 26
In vitro methods, 27–33
Irradiation, *see* Radiation
Isochromosomes, 59
Isozymes, lactate dehydrogenase and kidney tubule morphogenesis, 127

Jackson, H., 131
Jackson, W. P. U., 158
Jacob, F., 44
Janus green and soft part syndactyly, 108
Jost, A., 141, 145–146

Kalter, H., 20, 188, 193
Karnofsky, D. A., 20, 197
Kidney development, changes in isozyme pattern, 127
  effect of actinomycin, 127
  as a model-system, 31
  in *Sd* mutant, 89
  and viral susceptibility, 213
  in vitro, 31
Kilham rat virus (RV), 232
King, C. T. G., 207
Klein, N. H., 20
Klemetti, A., 17–18
Koskimies, O., 128

Landauer, W., 53–55, 190
Lash, J., 96, 100
Latent diseases and teratogenesis, 23
Lead salts and malformations, 123
Lens induction, 87
Lenz, W., 15, 199
Limb, effect of nitrogen mustard, 93–94
hereditary defects, *see* Mutants
inductive interactions, 90–94
in thalidomide embryopathy, 199, 203
Lloyd, J. B., 183
Loading test in inborn errors of metabolism, 43
Long-acting thyroid stimulator (LATS), 150–151
Lutwak-Mann, C., 26
Lyon hypothesis, 37, 70–71

McBride, W. G., 199
Maintenance factor in the limb, 91
Mammary gland, hormonal control of differentiation, 147
Maternal-fetal barriers, *see* Barriers
Maturation resistance, 212
Mechanical factors as etiology of congenital defects, 136–137
Meclizine, effect on human embryos, 205–206
in experimental animals, 207
prospective studies, 18, 206
structural formula, 205
Meclozine, *see* Meclizine
Mellin, G. W., 18, 206
Menkes, B., 108
Metabolic pathway(s), blocks, 41, 42
thresholds, 56
Metabolic inhibitors as teratogens, 183, 184
Microcephaly, and cytomegalovirus infection, 225
in irradiated human embryos, 178
Micromelia, drug-induced in chicken, 189
induction by aminonicotinamide, 53–55
Mitomycin C, and chromosomal aberrations, 62
Mitotic cycle, *see* Cell cycle
Mongolism, *see* Down's syndrome
Monod, J., 44
Moore, J. A., 96
Morphogenetic movements, in formation of heart, 79
in formation of secondary palate, 82
gastrulation, 79
Mosaicism, autosomal, 69
sex chromosomes, 71
Mouse strains, in analysis of cleft palate, 83
difference in sensitive periods, 126
differential susceptibility to 6-aminonicotinomide, 191, 192
differential susceptibility to galactoflavin and cortisone, 193
and induced cleft palate, 191
and induced vertebral anomalies, 191
susceptibility to teratogens, 20
Müllerian ducts, 145, 146
Mumps, and congenital defects in man, 225, 226
and experimental cataract, 229
and fetal wastage in man, 225, 226
Murakami, U., 121
Murphy, L., 183
Mutagens, 63
Mutant, anophthalmic (mouse), 88
brachydactyly (rabbit), 105
ch-hydrocephaly (mouse), 99
ch-Lamoreux' chondrodystrophy (chick), 54
crooked neck dwarf (chick), 107
lu–luxoid (mouse), 51
lx–luxate (mouse), 51
mm^A–Californian micromelia (chick) 54
mm^H–Massachusetts micromelia (chick), 54
polydactylous (chick), 91
rumpless (chick), 106
Sd–tailless (mouse), 89
$talpid^3$ (chick), 105
t^{w1} (mouse), 105
t^{w18} (mouse), 99
walzer-shaker (mouse), 105
wingless (chick), 91
Myeloschisis, due to overgrowth, 99
Myxoviruses, 215

Naeye, R. L., 222
Nakai, J., 33
Neel, J. V., 35
Nelson, M. M., 120
Neoplasia and congenital defects, 102
Nephroblastoma, 102
Nephrosis, congenital, 37, 38
Nerve growth factor, 97
Newcastle virus, spread in chick embryo, 230
Nitrogen mustard, production of phocomelia, 93
Nuclear detonation, *see* Atom bomb
Nowack, E., 202
Nucleic acid, inhibition of synthesis, 196–197

Nutritional conditions and teratogenic response, 23

Operon, 44, 45
Organ culture, advantages, 28
bone development, 28–29
cartilage formation, 96
closure of palatal shelves, 29–30
kidney development, 31
limitations and failings, 32
mammary gland differentiation, 146
methods, 28
muscle development, 32–33
pancreatic differentiation, 133
sex differentiation, 147
Organ specificity of sensitive periods, 121–122
Overripeness of eggs, 115
Ovulation, delayed, 116

Palatal shelves, closure, 82–85
fusion *in vitro*, 30, 85
strain differences in closure, 83
Pancreatic rudiment, actinomycin sensitivity, 133
differentiation, 133
Parrot beak, drug-induced in chicken, 190
Patten, B. M., 99
Phenocopy, 51–57
Phenylketonuria, 42–43
Phocomelia, experimental production in chick, 92–94
in thalidomide embryopathy, 199, 200
Pituitary, *see* Hypophysis
Placenta, endocrine function, 141–143
infectivity by viruses, 229
permeability to hormones, 144
permeability to immunoglobulins, 73
permeability to viruses, 228
Placental lactogenic hormone, 142
Polydactyly, in human population, 46, 47
induction by fluorouracil, 53
and inductive interactions, 90–91
polygenic control, 53
in *talpid*^{3} mutant, 107–108
Polyoma virus, different lesions caused by, 235
effect on metanephrogenic mesenchyme, 213
Polyploidy, 58
Pregnancy outcome, examination, 23–26
implantation sites, 24
screening methods, 24
Price, D., 146
Prospective studies, 16–18

Primary induction, 85–87
Prolactin, in mammary gland development, 147–148
Pseudohermaphroditism, in adrenogenital syndrome, 152
female, 151
as a result of androgen treatment during pregnancy, 155
Pteroylglutamic acid, sensitive period, 120
Puck, T. T., 166
Puromycin and chromosomal aberrations, 63

Radiation, cancerogenic effect, 179
and chromosomal aberrations, 61
and chromosomal aberrations in early zygote, 174–176
diagnostic, 178
at different stages of cellular life cycle, 167–170
effect on behavior, 176–177
effect in cell culture, 166–170
effect on cell proliferation, 168–169
effect on chromosomes, 61, 169–170
effect on DNA, 170
effect on early zygote, 174–176
effect on human development, 177–178
effect on sex chromosomes, 174–175
injury at cellular level, 166
and leukemia, 179
lethal effect on mouse embryo, 171, 172
low level, effect of, 178–179
from nuclear detonations, 178
and nucleic acid precursors, 170
repair of cellular damage, 170–171
and reproductive death, 166–167
sensitive periods, 124, 134, 172–173
sources of, 178
teratogenic action, 171–174
types, 165
units, 166
Radioactive isotopes, $-I^{131}$, effect on fetal thyroid gland, 150
as sources of fetal radiation, 178
Reaggregation of embryonic cells, 129
Registers of congenital defects, 8
Reproductive tract, control of differentiation, 145–147
Restoration capacity, 128–130
Restricted diet, effect on fetus, 188
Retrospective studies, memory bias, 17
Rhesus antigen, 73–74
Robertson, G. G., 229
RNA transcription, 44, 113

Rous sarcoma virus, different lesions caused by, 235
Rubella, and chromosomal abnormalities, 66
epidemiology, 14
fetal wastage, 217, 226
inhibition of cell multiplication, 98
persistent inflammation, 219
risk of congenital defects, 217–218
route of infection, 221
sensitive period, 122, 217
syndrome, *see* Embryopathy
Rubella embryopathy, clinical picture and morbid anatomy, 219, 220
pathogenesis, 219
Runner, M. N., 188
Russell, L. B., 113, 124, 174
Russell, W. L., 113, 124
Rutter, W. J., 132
Röntgen, W. K., 165

Salicylamide, effect on mucopolysaccharide synthesis, 198
Salzgeber, B., 94
Saunders, J. W., Jr., 90, 104, 108
Saxén, L., 15, 29, 198
Seasonal variations in teratogenic response, 23
Sensitive periods, to Actinomycin D, 119, 127, 198
of axial skeleton malformations, 121
and barriers, 130
determining factors, 126–134
in different mouse strains, 126
of hemoglobin synthesis to different metabolic inhibitors, 123
and intensity of treatment, 125, 134
interstrain differences, 125–126
in man, 114
in mouse, 114
and nucleic acid metabolism, 133, 197, 198
to nutritional deficiency, 120
organ specificity, 121, 124
and proliferation of cells, 131, 168
to radiation, 113, 124, 167–170, 175
in relation to stage of development, 114, 230
and restoration capacity, 128–130
in rubella embryopathy, 122, 217–219
to short-term hypoxia, 121
in thalidomide embryopathy, 202
Sex chromatin, 71
Sex chromosomes, abnormalities, 69–71
Barr body, 71
loss after radiation, 174–175
Sex chromosomes (*Continued*)
Lyon hypothesis, 70–71
mosaicism, 71
Sex differentiation, hormonal regulation, 145–147
two principle theory, 146
Sex reversal, caused by hormones, 145
Screening method for embryos, 24–26
Shear, H. H., 216
Shope papilloma virus, 232
Sickle-cell anemia, 74
Siegel, M., 226
Smallpox and vaccinia, fetal wastage, 224
Somatotropin, effect on pituitary dwarfism, 149
Spherocytosis, hereditary, 46
Statten, P., 14
Steroid hormones, metabolism in fetoplacental unit, 142–143
Steroid metabolism, congenital disorders related to, 151–152
Stillbirths, incidence of defects, 5
Stockdale, F. E., 147
Sulphanilamide, combined action with aminonicotinamide, 189
Syndactyly, production with Janus green, 108
Synergistic action, of genes and exogenous factors, 51–57
of several teratogens, 187–191

Talipes, 137
Taylor, R. J., 9
Teratoma(s), 102
Testis, effect on sex differentiation, 145–146
Testosterone, effect on fetus, 155
in experimental adrenogenital syndrome, 154
Tetracycline, effect on bones *in vitro,* 29
effect on the mother, 23
effect on teeth, 136
inhibition of calcification, 136
in the skeleton of newborn children, 136
Tetralogy of Fallot, 100–101
Thalidomide, metabolism, 204
metabolites, 204
and panthothenic acid deficiency in rats, 204
in rabbit semen, 114
structural formula, 199
Thalidomide embryopathy, in experimental animals, 203–204
in man, 199
in monkeys, 203

Thalidomide embryopathy (*Continued*)
pathogenesis, 203–204
sensitive period in man, 202
and wholesale figures, 15, 201
Thiouracil, effect on fetus, 149
Thyroid gland, defects of development, 149–150
Thyroid hormones, in amphibian metamorphosis, 144
deficiency of, 150
Thyroid-stimulating hormone, overproduction of, 150
Thyroxine, placental transfer, 144
in treatment of pituitary dwarfism, 149
(*see also* Thyroid hormones)
Tissue culture, (*see also* Organ culture), 27
Tissue interactions, *see* Inductive tissue interactions
Tolmach, L. J., 169
Toxoplasmosis, congenital defects, 135
hydrocephalus, 135
Transplantation, intrauterine, 26
Triethylene-melamine (TEM) and anophthalmia in newt larvae, 194, 195
Trisomies, 66–67
Trophoblast, interaction with maternal organism, 73
Trypan blue, mode of teratogenic action, 186
teratogenic effect on rat embryo, 184, 185
TTSD of teratology, 19
Turner's syndrome, 155
Töndury, G., 194, 195, 219, 221

Vaccination against smallpox during pregnancy, 9, 224
Vainio, T., 213
Vertical transmission of virus infection, 230
Viremia, in different viral infections, 227
Virilization, in adrenogenital syndrome, 152
as a result of hormone treatment during pregnancy, 155
Viral infection, and congenital defects in man, 222–226
and fetal deaths in man, 226
methods of detection in mother and offspring, 222
and viremia, 227
Viral lesions, different types of, 211, 235
Viral susceptibility, and cell proliferation, 231
Viral susceptibility (*Continued*)
of embryonic cells, 211
and endocrine factors, 234
factors affecting, 227–235
and metabolic activity of cells, 232
and production of interferons, 232
Virus, avian leucosis, 230
blue tongue vaccine and fetal wastage in cattle, 216
and chromosomal aberrations, 63–66
coxsackie, 226
cytomegalo, 224
echo, 225
effect on host cell, 211
H-1 and malformation in hamster, 217
hog cholera and malformations in pig, 216
and host cell metabolism, 211
and inductive interaction, 213
influenza, 215, 223, 226
interaction with cells, 210
and malformations in experimental animals, 215–217
and malformations in man, 222–226
mumps, 225
myxoma, 212
myxoviruses in embryonic chicks, 215
oncogenic effect and maturation of immune system, 234
polyoma, 213
polyoma in kidney organ cultures, 213
receptors in embryonic cells, 231
rubella, 217–222
smallpox, 214
vertical transmission from mother to offspring, 230
(*see also* specific viruses)
Vitamin A, effect on hamster embryo, 194
effect on mucopolysaccharide synthesis, 198
and viral susceptibility, 214

Waddington, C. H., 51, 112
Warkany, J., 2, 21
Wells, L. J., 141
Willis, R. A., 106
Wilms' tumor, *see* Nephroblastoma
Wilson, J. G., 21, 119
Wilt, F., 123
Witschi, E., 115
Wolff, Et. 26, 91–93
Wolffian ducts, 145
Wood, M., 14

X chromsome, *see* Sex chromsome

Zwilling, E., 90–92, 104, 106

OTHER BOOKS OF INTEREST

Molecular and Cellular Biology Series

THE REGULATION OF CELL METABOLISM

Georges N. Cohen, Directeur du Laboratoire d'Enzymologie, Centre National de la Recherche Scientifique, France

STRUCTURAL CONCEPTS IN IMMUNOLOGY AND IMMUNOCHEMISTRY

Elvin A. Kabat, Columbia University

HOLT, RINEHART AND WINSTON, INC.
383 Madison Ave., New York, New York 10017